History of Technology

History of Technology

Fifth Annual Volume, 1980

Edited by

A. RUPERT HALL and NORMAN SMITH

Imperial College, London

MANSELL PUBLISHING

ISBN 0 7201 1585 X

ISSN 0307-5451

Mansell Publishing, a member of Bemrose U.K. Limited,
3 Bloomsbury Place, London WC1A 20A

First published 1980

British Library Cataloguing in Publication Data

History of technology.
  5th annual volume: 1980
  1. Technology — History — Periodicals
  I. Hall, Alfred Rupert II. Smith, Norman
  609   T15

ISBN 0-7201-1585-X

Typeset by
Preface Ltd., Salisbury, Wiltshire
Printed in Great Britain by
The Scolar Press, Ilkley, West Yorkshire

Contents

Preface

What is the core of the history of technology as a discipline? Some would suppose it to be the study of the way machines, structures and processes have been designed, fabricated and developed: in the last resort, in this view, it is the historian's task to define the nature of an engineering or production problem as it was faced in the past, and then analyse the ideas and techniques which were used to achieve a solution. Ultimately the result should be something more than mere compilation of facts and figures, data and dimensions. The greater purpose is to perceive something of engineering's methodology, its intellectual content and the nature of the engineer's ideas and attitudes to problem solving. We have taken such an idea of the history of technology as the core of our studies in these annual volumes, as of our teaching at Imperial College. The history of technology can properly stretch — dare one hope in a normal distribution curve? — from the recording and description of 'industrial archaeology' at one extreme to social analysis at the other.

This year we have, in fact, been even more elastic than before. We print for the first time some pages of industrial revolution description; we offer some comparison of a 'primitive' modern technology (in paper-making) with that of the past; but at the other extreme we also publish some critical reflections upon systems in the development of technology, and a note on Galileo and measurement (a paper on Galileo and the development of modern technology would be very interesting too). In between are papers on pumps and harvesters nearer to our mean position.

A. RUPERT HALL
NORMAN A. F. SMITH

The Order of the Technological World*

THOMAS P. HUGHES

As late as the nineteenth century, western technology was still thought of as harnessing the powers of nature. It was seen as creating a habitable world within a vast natural context. Great Britain, the most technological of nations, was in process of rapid industrialization and urbanization, although the majority of the population remained rural and lived in intimate association with nature. Man was not eliminating nature but improving upon it, as canals replaced rivers, tunnels supplemented natural passes, machines substituted for labour, and steam engines replaced animals, wind, and water power. Gaslight was extending the day, improved heating was moderating the effects of climate, and the early telegraph lines were eliminating one aspect of space. Nevertheless, nature was never more than a horizon away and her great forces still determined abundance or famine, health or disease, comfort or misery.

Now, western technology is the environment and the great determining force in the modern industrial world. This generalization has become a truism, but its ramifications are only beginning to be comprehended. The majority of the population lives in a man-made environment where nature is only a tree, a patch of blue sky, a short exposure to excessive heat, cold, or rain. Natural rhythms have given way to man-made schedules. Natural substances such as wood are much less in evidence; even water and air are full of man-made substances. This knowledge is commonplace, a cause of considerable lament and even anxiety, but analysis, understanding, detailed description of the man-made environment are sporadic, disjointed, and generally unenlightening.

Historians, especially historians of technology, have an extremely important key to understanding the new world, for they have the potential to describe and explain the technological structures ordering the modern world. Students of natural evolution have done much to enlighten us about the nature and the dynamics of natural structures. Historians, however, have as yet only scratched the surface of the highly organized and evolving man-made world. They have not explained the evolution of such omnipresent systems as streets and

*Revision of a paper first read in August 1978 at a conference on 'Critical Problems in the History of Technology' sponsored by the National Science Foundation.

highways, telephones and television, pipelines and electric transmission grids, computer networks and airline routes. Nor have they satisfactorily explained the dynamics of manufacturing systems, chemical processes, and vertically integrated, technologically structured, international enterprises. There are many reasons for these failures; for one, the man-made systems are more complex than the Newtonian and apparently more difficult to comprehend than the Darwinian. A Newton or Darwin for the man-made world is wanting.

This leads us to the question of definition. What are the systems that should be explained by the historian? What are these structures that increasingly determine the history of the world? First, it should be established that they are old. Second, it should be stated that they tend to become larger and more complex with the passage of the centuries. Thirdly, it should be noted that formerly the largest ones evolved, but now more and more and larger and larger ones are designed, constructed, and managed by man. Having had success in planning and constructing small systems like the machines, then larger ones like the factory, man in the twentieth century has drawn upon his empirical knowledge of these to try such monstrous systematic endeavours as the Manhattan project.

Historians dealing with broad sweeps of time and widely separated places need be wary of narrow definitions. Engineers, scientists, and social scientists focus their attention more sharply, so they can define more precisely. The historian's definition of system often has to take into account the varied ways in which the term has been used in the past and the many variations of usage among his professional peers. The entry for 'system' in the *Oxford English Dictionary* is a reminder that no single definition will encompass the varied subject matter of the historian. Ludwig von Bertalanffy needed a book, not a paragraph, to define 'system'.[1]

Nevertheless, there are some characteristics of systems so general that a statement about their essence is possible. A system is constituted by related parts, or components. Because the parts are related, the state, or activity, of one influences the state, or activity, of others in the system. For example, in a system (as described below) integrating production and distribution, changes in distribution (or production) send waves of change through production (or distribution). A system is not simply a sum or a heap. For instance, in the Lynn White system described later, the plough, animal team and ploughman together are a system; a number of ploughmen, teams and ploughs working in the same vicinity would be a sum or a heap. A system has characteristics different from those of its components in isolation. In an electric power system this is manifest. Control of a system is often centralized and man often closes the loop. The managers in the Chandler systems (analysed below) are examples. The environment of a system is those things and people 'that are "fixed" or "given" from the system's point of view'.[2] (Many system builders in history have found the environment an irritant.) Systems either evolve by confluences, or they are planned and designed. Historians encounter extremely complex systems that originate and grow in both ways. The system of production described by Marx and prevalent during the British Industrial Revolution resulted from some planning and many more confluences of an accidental kind. It should be observed that the historian usually encounters technological systems that are open, not closed. If the systems were closed, their final state could be predicted from their initial condition, and historians rarely fail to belabour a point about predictability.

With these essentials of the system as points of departure, the following essay selects several major historical works to suggest how their writers have used the concept of the system to organize, analyse and draw conclusions about the history of technology from disparate materials. Innumerable other histories of significance could be used to sustain the argument. The recent books and articles of Louis Hunter, Otto Mayr, Nathan Rosenberg, Edward Constant, James Brittain, Robert Belfield, Hugh Aitken, J.F. Hanieski, David Noble, Langdon Winner, Thomas Smith, David Hounshell, M. Roe Smith, David Landes, Elting Morison, Arthur Johnson, Russell Fries, and Reese Jenkins come immediately to mind.[3]

Deep Ploughs, Harnessed Teams and Communes

In 1962 Lynn White eloquently described the emergence of a system of agriculture in the early Middle Ages on the plains of northern Europe and England. He called the system the 'agricultural revolution of the early Middle Ages'.[4] The northern medieval peasants worked out a 'new system of food production more balanced and efficient than anything earlier',[5] White concluded (in an article published in 1967), and added, 'it was an agricultural revolution unparalleled since the first invention of tillage. Its elements — the heavy plough, open fields, three-field rotation, and horse harness — accumulated and consolidated into a new agrarian system from the 6th through the 9th centuries.'[6] White's use of the world 'consolidation' suggests his emphasis upon system; if doubt remains, then consider his writing of the 'interlocking' of a new pattern of cereal growing (open field and three-field) with improvements in cattle raising (the village herd could graze and leave their droppings on the fallow field of the three-field system).[7]

A system's components interlock so that there is an interdependence. A system changes the functioning of its components from what it would be — if they existed at all — outside the system. In the case of White's agricultural system, the heavy plough with wheels, coulter, horizontal ploughshare, and a mouldboard was a leading component and affected the characteristics of the other components in the system. The heavy alluvial soil of the northern European plain was an environment requiring a different plough from the light, two-ox scratch-plough commonly used earlier on the light soils of the Roman peninsula. The heavy plough came to the Slavs from unknown sources and then its use spread to others through the Frankish heartland in the seventh century. They found that friction of the new plough cutting through the heavy soil was so great that an eight- — not a two- — oxen team was needed. With the harnessing of the large team to the plough a coherent system of agriculture emerged. In contrast, 'no coherent new system of cultivation' had emerged in Roman times in the northern provinces.[8]

The internal logic of the evolving system also became manifest in the resort to the communal social structure. White reasoned that an eight-oxen team represented such a substantial capital accumulation that individual peasants had to pool their resources. They combined their strips into open fields and engaged in a common, scheduled endeavour, necessitating decisions about time of planting and harvesting that were binding upon the collective. White wrote, 'The adoption of the new plow therefore helps to explain the communal pattern of manorial life in Northern Europe.'[9]

Some medieval person, or persons — because simultaneity of invention is common — conceived of the system of heavy plough and ox team. The eureka moment may have seemed no more dramatic to the inventor than the idea of adding more animals because the two-oxen team was not adequate, but equally likely is the sudden and embracing idea of making a heavier plough drawn by a larger team in order to plough the alluvial soils. In the latter case, a system was invented. Perhaps Lynn White or one of his students will some day miraculously name the inventor, or inventors, for White has been able to name an inventive medieval flyer.

The coherent system of cultivation was an open one. It not only spread through northern Europe and into England with the Danes, but it also evolved. Consequences of the new system, unforeseen by its inventors and users, implied changes in the system so that it might fulfil more effectively its mission of food production. In the later eighth century, peasants in the region between the Loire and Rhine rivers, where the summer rains were adequate, began the three-field system of cultivation. This stride may have been taken in parallel with the introduction of the heavy plough and the eight-oxen team. The parallel developments, however, interacted. The summer-harvested crop of the three-field system often included oats, a cereal not common under the two-field system. The availability of oats, which allowed the peasants to keep horses, and the superior characteristics of the horses in the plough team soon brought about their increased use.[10] Horses not only increased agricultural yield, but were bred for the mounted warrior of the feudal system. By the twelfth century, then, the agricultural system included not only the plough, the multi-animal team, and the commune, but a team of horses rather than oxen. Lefebvre des Noëttes has explained the introduction of the more effective collar harness for the horse whose physiognomy did not suit the yoke harness of the oxen.[11] Moreover, the horse was shod with iron and this increased his efficiency as a motive power. Shod horses with collars became an integral part of the evolving system in many parts of Europe.

To define the borders of a system is virtually impossible; to limit the analysis of dynamic interactions of the agricultural revolution to agriculture alone simplifies the problem of history writing, but does not do justice to its complexity. Marc Bloch, White and others have found that the new agriculture interacted with other momentous medieval developments such as population growth, urbanization, increasing leisure, and the rise of 'democratic capitalism'.[12]

Labour, Machines and Automatons

Karl Marx's chapter in *Capital* entitled 'Machinery and Modern Industry', is an extended analysis of the machine system of production. In explaining how the machine system contrasts with the earlier production systems of handicraft and manufacture, Marx offers a history of the evolution of the machine system. He further defines the machine system by explaining how its components, parts, or subsystems interact and how they became in their most highly developed state in the mid-nineteenth century a great automaton. In a lengthy discourse, he stresses, as many other historians of technology do not, the role of labour as a component in the machine system of production.

Marx lucidly defined the modern industrial system of machine production in a brief passage:

Each detail machine supplies raw material to the machine next in order; and since they are all working at the same time, the product is always going through the various stages of its fabrication, and is also constantly in a state of transition, from one phase to another. ... In an organised system of machinery, where one detail machine is constantly kept employed by another, a fixed relation is established between their numbers, their size, and their speed. The collective machine, now an organised system of various kinds of single machines, and of groups of single machines, becomes more and more perfect, the more the process as a whole becomes a continuous one, i.e., the less the raw material is interrupted in its passage from its first phase to its last; in other words, the more its passage from one phase to another is effected, not by the hand of man, but by the machinery itself.[13]

Notably the process of flow (or 'through-put', to use recent jargon) is described without reference to human labour, for it, as explained (see below), is a mere component whose function is determined by the components of the machine system.

Marx further clarified the nature of the machine system by contrasting it with handicraft and manufacture. He used carriage-making as an illustration. In the era of the craftsman, the carriage was the product of the labour of a large number of independent artisans, such as wheelwrights, fringe-makers and gilders. In manufacture, however, the different artisans were, according to Marx, 'assembled in one building, where they work into one another's hands'.[14] As a result cooperation was greater and a system more like the modern industrial machine system emerged. A notable transformation took place in manufacture when a drive to increase production brought functional specialization of the craftsmen. Their function was then determined by the logic of carriage production; they became, under advanced manufacture, functionally specialized artisans labouring within a factory space given over entirely to carriage manufacture. A system of manufacture had developed.

Marx also showed how manufacture arose as a result of a capitalist's assembling or accumulating in one place, or manufactory, a number of artisans labouring side by side and in precisely the same way. For instance, a number of craftsmen were gathered and each made an entire commodity such as paper or needles. In time, however, the logic of production brought a redefinition of labour and the subdivision of the process of making the commodity into different and sequentially related tasks. Then the paper or needle factory housed a system of manufacture like the carriage system described above.[15]

The system of manufacture was not the machine system. Manufacture evolved into a machine system as machines replaced tools in the hands of the artisans. In describing the machine system — the focus of *Capital* — Marx repeatedly used the language now associated with systems engineers and managers. Not only did he stress flow, but also the imperative, or determining force, of the system of machinery. In manufacture the division of labour organized the process; in the machine system, the machines. He quoted Andrew Ure, who wrote that the principle of the new system was 'to substitute . . . the partition of the process into its essential constituents, for the division or graduation of labour among artisans'.[16] The result was an 'organised system of machinery in a factory';[17] 'machinery organised into a system';[18] and 'each machine constituting a special organ, with a special function; in the system'. Like Charles Babbage, author of an *Economy of Manufactures and Machinery* (1832) before him, Marx finally called the machine system an automaton. This he did especially when all the machines with their various and specialised tools were interconnected by shafts, belts, and gears to a single steam engine. A system of machinery, he observed, 'constitutes in itself a huge automaton, whenever it is driven by a self-acting prime mover'.[19] Marx realized that the 'self-acting' steam engine with its governing mechanism implied an automaton,

'The most developed form of production by machinery.'[20] He spoke of 'a mechanical monster whose body fills whole factories, and whose demon power, at first veiled under the slow and measured motion of his giant limbs, at length breaks out into the fast and furious whirl of his countless working organs'.[21] The late twentieth century prefers to refer to automation.

Marx respected the intellect and organizing creativity embodied in the machine system of production, but he also lamented — to a far greater extent than most, who comprehended but who were outside the system — labour's function in the system. Labour he saw as a mere appendage of the machine system. The labourers had become organs, 'co-ordinate with the unconscious organs of the automaton, and together with them, subordinated to the central moving power'.[22] Labour was merely an object. Sensitive to all concerns of labour, it is not surprising that Marx resorted to metaphors suggesting that the system was a fast and furiously whirling beast.

This is not the place to explore in detail Marx's penetrating and detailed analysis of the impact of the machine upon the worker. Other insights of his into the systems of the machine age can only be mentioned. The machine systems of production related to one another, Marx believed, as subsystems in a massive, overarching production system spreading throughout industrial society. He realized that a radical change in the mode of production in one sphere of industry caused reverberations and coordinated changes in another. The introduction of the machine system in spinning, for instance, impelled machine weaving. Machine spinning and weaving suggested, even necessitated, the rationalization of chemical processes for the production of bleaches and dyes. The nature of these evolving and interacting systems implied the mechanization of transportation and communication networks to bind and inform.[23] As he expanded his horizons of analysis, technological determinism as a major explanation of change in history emerged. As he focused upon the capitalists and their institutions, Marx saw the driving force of systematization and centralization that was organizing into a coordinated system all of industrial society, its relations of production, and its superstructure.

Working the Visible Hand

Alfred Chandler, a professor of history at the Harvard Business School, also sees history as evolving systems. Marx's grand scheme of interpretation encompassed all human activity in technologically structured systems; Chandler's horizons do not extend as far but, within his field of vision, his delineation of systems is richer and more precise. Chandler extends his systems analysis beyond production to distribution and encompasses them both. He investigates in depth the infrastructure upon which the linked production and distribution systems depend. Chandler, however, focuses not upon labour as a component in the system — it is scarcely mentioned — but upon the 'visible hand' of middle managers who coordinate and monitor large systems. Modern industrial production, Chandler observes, is capital, energy, and manager-intensive — not labour-intensive. Like Karl Marx, however, Chandler writes as a technological determinist in his recent Pulitzer Prize-winning study, *The Visible Hand: The Managerial Revolution in American Business*.

Chandler's history of expanding systems covers a century and a half. The last part deals with the 'Modern Industrial Enterprise'. Along the way, he treats the 'revolution in transportation and communication', 'the revolution in distribution and production', and 'the integration of mass production and mass distribution'. His key concepts throughout the development of his major thesis and its sub-themes are 'coordination', 'cooperation', 'control', 'throughput', 'flow', 'integration', 'vertical integration', 'concentration', 'function', 'structure', and 'management'. This is the language of systems and reminiscent of, if richer than, Marx's vocabulary.

He also describes the evolution of large systems and their transformation of previously existing systems into subsystems. He begins his analysis of change with the textile factory and machine system of Marx. America (according to Chandler), before England, witnessed the appearance of the integrated spinning and weaving factory, first at Waltham in 1814 and then at Lowell, Massachusetts. The integrated textile mills at Lowell were the largest industrial establishment of their day. The integration of all the processes of textile production stimulated 'innovation in each of the specific processes'.[24] The rapid and complex flow of materials presented challenging problems of coordination and monitoring, problems that would also bring the visible hand into other industries but several decades later.

A necessary but not sufficient cause of the emergence of the modern industrial enterprise was the full development of an infrastructure consisting of telegraph network, national railway system, and, later, an extended pipeline system. Chandler writes, 'as the basic infrastructure came into being between the 1850s and 1880s, modern methods of mass production and distribution and the modern business enterprises that managed them made their appearance'.[25] The infrastructure made possible the linking of production systems, which had only been loosely coordinated by the market, into encompassing systems presided over by single enterprises imposing integration and close coordination and allowing a 'new velocity of output and flows'.[26] The flow of information over the telegraph and later the telephone network was as necessary to development as the movement over the standardized and linked national railway system, which had succeeded canals. Subsequently highways and airlines extended and intensified the infrastructure, but Chandler directs his attention primarily to the railways.

A critical stage in the Chandler history is the transformation of production facilities along lines adumbrated at Lowell. The measure of the transformation was accelerated 'throughput' or flow. Ingeniously, Chandler identifies those spheres of production where the materials are themselves fluid and the energy usage is high as those most responsive to the drive for increased production through coordination, concentration and control. He singles out the furnace and foundry and the distilling and refining industries.[27] In other industries, such as metal-working and the mechanical, increased throughput[28] came more slowly because technological and organizational innovations were more demanding. As an example of technological innovation clearing the way for flow, he suggests materials-handling equipment; as an example of organisational innovation he refers to Frederick W. Taylor's scientific management.[29] At a later date the 'coordination of throughput from the crude oil wells through the pipelines to the refineries' is cited as an example of the systematization of production.[30]

If the production systems only had accelerated flow, then horrendous blockages at the distribution end would have occurred. So, not surprisingly, Chandler finds that the modes of distribution were also coordinated, centralized, and controlled. Technology was the major means to develop the modern system of production; administrative and organizational reform played a comparable role in the systematization of distribution. 'By means of such administrative coordination,' Chandler writes, 'the new mass marketers reduced the number of transactions involved in the flow of goods, increased the speed and regularity of that flow, and so lowered costs and improved productivity of the American distribution system.'[31]

Until the late nineteenth century, the market usually coordinated the production and distribution systems, and the infrastructure provided linkages. Increased velocity and volume in each system, however, brought the organizers and systematizers to resort to vertical integration, a phenomenon familiar to economic and business historians, and soundly integrated into his explanatory scheme by Chandler. Vertical integration allowed coordination, centralization and control of the two systems and the creation of an overarching system of production and distribution under one enterprise and its management. Chandler cites numerous examples of the phenomenon.[32] Within three decades after 1870, these integrated enterprises 'came to dominate many of the nation's most vital industries'. These were the first 'big businesses' in America; they were the industrial enterprises that were 'the archetype of today's giant corporation . . . the integration of the processes of mass production with those of mass distribution within a single business firm'.[33]

But the capstone has not been laid — the 'visible hand' must be seen. The man at the heart of the system, the person who knew its innermost workings, is the modern professional manager.[34] The manager, the visible hand, 'monitors the processes of production and distribution' and 'coordinates the high speed, high volume flows through them more efficiently than if the monitoring and coordinating had been left to market mechanisms'.[35] (This view has caused raised eyebrows in those places that cherish a market system — or at least the idea of one.) So, the visible hand of management is replacing the invisible one of market forces. There is no reason to assume that Marx extrapolated would contradict this conclusion.

Chandler, too, is a not unfamiliar presence in the camp of the technological determinists. He is too experienced and complex a historian to write of simplistic determination, but he does see and stress that technology was at the root of the emergence of the modern systems of production. And he gives this prominent place to technology despite having defined it narrowly as innovations in 'materials, power sources, machinery, and other artifacts'.[36] He defines as organizational innovation the way in which the artefacts — as well as the workers — are coordinated, controlled and arranged. Furthermore, he explains success in vertical integration as a function of the basic technology of the industry. The technology facilitating mass production and systematization supported vertical integration on the organizational level. Chandler goes so far as to say that 'technology of production was certainly the critical determinant in the growth' of the integrated firm.[37] Furthermore, he characterizes organization or institution changes as technologically determined when he writes, 'the modern factory was as much the specific organizational response to the needs of the new production technology as the railroad and the telegraph enterprises were responses to the operational needs of the new technologies of transportation and communication, and as the mass marketing firm was to the opportunities created by those same technological advances'.[38]

Power Systems

Chandler stresses the essential flow, coordination and centralization of modern industrial systems. He writes about systems in many spheres of industry, but he omits the paragon of early twentieth-century systems — electric power. In the long history of technology, the span of years from about 1880 to 1930 is clearly a half-century well characterized technologically by the emergence and development of city centre, urban district, and then regional power systems. To many Americans, the construction of the Tennessee Valley Authority regional system in the 1930s was the epitome of electric light and power developments, but elsewhere in the world by then there were other, comparable achievements. Today regional power systems are commonplace, and the United Kingdom and the Soviet Union have national power systems.[39]

In the electric power system, flow, or throughput, reaches an ultimate for the flow is at the speed of light, a state not even approached by Henry Ford's fast-moving assembly lines at River Rouge. Supply and demand are so linked in a power system that change in one in relation to the other is instantaneous. The manager of an integrated railway system had to monitor closely the demand for transportation and the location of his rolling stock so that he might respond sensitively, but the load dispatcher in a light and power system sees by the flick of a needle indicator or the count of a computer display the switching on of lamps during a summer storm at the same instant as the consumer sees his lamps light up. Generally, the commodity cannot be stored; electric power must be consumed as it is generated.

Chandler explains how flow was accompanied by managerial development of coordination and centralization. The speed-of-light flow of electricity demanded the introduction of new managerial responses. It was in connection with the rise of power systems that managerial concepts such as diversity and load factor were articulated and perfected. A century ago Karl Marx discussed the essence of load factor, but electrical engineers and managers defined it. Marx discourses at length to explain how the tyranny of capital investment drove the factory owner to work his machines — and the attendant labour — longer hours and more intensively in order to exploit fully his investment. The manager of an electric light and power system drives his machines to the ultimate, too. In so doing he measures his success by the value of his load factor, or the shape of the load curve that graphically represents it. Labour does not concern the utility manager as much as the factory owner because electric power is a highly capital- — or machine- and equipment- — intensive industry. Much as the ideal of the nineteenth-century factory owner was a tireless twenty-four-hour worker, the ideal of the utility manager is a flat load curve showing that the installed capacity of generation is fully used around the clock.

The managers of railways and of vertically integrated industries may have kept load curves and refined the concept of load factor, but Chandler does not tell us this. Problems of managing utilities also brought the articulation and application of the concept of diversity. Managers of other industries and expounders of economic theory writing earlier also understood the diversity principle, but the electric light and power utilities refined and defined it. Samuel Insull, the innovative manager of Chicago's Commonwealth Edison Company, often lectured in the twenties to his peers and other businessmen about the diversity principle, a *sine qua non* of modern utility management.

Insull and other major utility managers conceived of both supply and demand as systems. In the case of demand, a high diversity of components, or loads, prevented concentration, or peakings, of demand that uniformity would have brought. A diverse mix of light, motor, electrochemical and traction loads usually levelled the demand curve. The managers also looked for a diverse and economic mix of generating components in the supply system, for at different times and in different places different kinds of energy, thermal and hydroelectric, were economically desirable. This emphasis upon diversity in systems is a reminder that large systems are not necessarily homogeneous or monolithic.

Another contribution of the evolving power systems to the general knowledge of systems was mathematical analysis of a highly effective kind. Chandler judges his modern managers as advanced, but none he describes write equations to describe the systems they managed. In the case of electric light and power systems, engineers and appliers of mathematics and science — like the remarkable Charles Proteus Steinmetz — wrote equations describing the relation of the measured and variable components in circuits and systems. After alternating and polyphase currents were introduced into electric power transmission, Steinmetz made a notable contribution by developing the relevant equations, which were far more complex than those applicable to the direct current systems of an earlier date. Armed with these equations, the engineer and manager rationalized and more effectively managed larger systems. One reason electrical technology is considered scientific and railway technology is not is the ability of the utilities so eloquently to describe their technology.

The modern power system is also a paragon of centralization. Karl Marx viewed with fascination the machine factory system as an automaton. Lynn White is not unmoved when he sees the components of his agricultural revolution fall neatly into place. How much more aesthetically satisfying is the contemporary regional power system with its infinitely discerning controls centred in, to use Marx's term, a self-acting computer. Energy control centres, as some utilities now style them, originated about the turn of the century in the load-dispatching centre where the dispatcher could manually switch, or connect and disconnect, the supply and load components. Now switches are remotely controlled, as are the generating units and other items of equipment.

The state of these supply side components and of the load centres is constantly monitored by telemetering or other information subsystems. The computer receives, processes, displays and stores the information about the state of the system and the actions of the load dispatcher at the master control point in the energy control centre. The computer knows the details about all of the system components and can, therefore, tell the engineer at the control centre what the results of his various actions may be. The computer, part analogue and part digital in some centres, is often programmed to take action to maintain the fundamental relationship between demand and supply. The computer can be programmed to increase the steam supply to turbines, for instance, if the system frequency, the critical indicator of the relation between supply and demand, should drop. Despite the casual use of words like automation, however, power outages resulting from human error in control centres remind us that man still manages the system. The ploughman had to intervene, or close the loop, and drive his team harder when the earth to be tilled became dense and moist; the load dispatcher performs analogously.

Conclusion

Only a few systems of technology could be discussed in this essay, but the range of these suggests how widely the concept of systems applies. The agricultural revolution occurred in the Middle Ages in a pre-industrial society; Marx wrote of a rapidly industrializing nation; and electric power systems are contemporary. Historians of technology can identify numerous other systems in the history of technology. Sometimes consciously, often unconsciously, they use 'system' to name that about which they are writing. A few years ago, a two-volume survey of western technology appeared in which the authors often organized their essays by using the systems concept.[40]

Despite the widespread use of the concept, only a few historians, such as Chandler, have explicitly explored the structure and the evolution of the systems they identify. Chandler, by covering the broad sweep of American history, encompasses the structure and operation of extended systems. His explanation for the evolution of systems is eloquent and persuasive, but abstract. In *The Visible Hand*, managers are most prominent, while engineers receive scant attention, and inventors the same. Historians focusing upon technology and using systems as an organizing concept have an especial responsibility to pay more heed to the inventors, engineers and scientists than does Chandler. Similarly, Marx's preoccupation with the capitalist obscured his vision of systems evolving. If the engineers and inventors are introduced, then the invention, development and rationalization of systems will become a part of their history. History of technology as evolving systems should introduce sociological analysis as well.

Historians who focus upon systems, however, should not limit themselves to a structural analysis; they should develop explanations for system dynamics. This will require, among many other analytical tools, a way of comprehending the changes in an open system. For example, the historian should be able to describe persuasively the shift from an urban-centred power and light system to a regional one. The inertia of highly developed systems also begs explanation. Notions about system imbalances, bottlenecks, reverse salients, and critical problems need refinement. The problem-solving techniques and the psychology of systems builders, be they inventors, engineers, scientists or managers, await further investigation.

Finally, reference should be made to everyman's anxiety about systems. As this paper has suggested, the intensity, scope, and organizing power of systems have increased dramatically over the centuries. Persons who have no interest in the theory or history of systems feel overpowered and manipulated by systems today, hence the derogatory reference to 'the system'. Often reference is to political and bureaucratic systems ('the establishment' in English terms), but as an awareness of the nature of modern technology increases, the public consciousness of technological systems grows. Associated with systems in the minds of many is the man who lives by the systems, 'the technocrat'. The use of the term in popular literature is often vague and the associations, such as those Theodore Roszak makes between technocrats and systems, are sometimes tenuous.[41]

Nevertheless, this essay shows that concern, even anxiety, about the power of systems in our lives is reasonable. From the Middle Ages to the present, the makers and managers of technological systems have been able to increase flow, coordination and centralization. The systematizers have, our cases suggest, repeatedly enlarged the boundaries of the systems and reduced the extent of the environment. We are reminded that people and things in a system are controlled; what is beyond control is the environment. There is an ironic double entendre in the call to save our environment.

Notes

1. Ludwig von Bertalanffy, *General Systems Theory: Foundations, Development, Applications* (New York, N.Y., 1968).

2. C. West Churchman, *The Systems Approach* (New York, N.Y., 1968), p.35.

3. H. Aitken, *Syntony and Spark: The Origins of Radio* (New York, 1976).
Robert Belfield, 'The Niagara System: The Evolution of an Electric Power Complex at Niagara Falls, 1883–1896', *Proceedings of the IEEE*, LXIV (1976), 1344–50.
James Brittain, 'The Introduction of the Loading Coil: George A. Campbell and Michael I. Pupin', *Technology and Culture*, XI (1970), 36–57.
E.W. Constant, 'A Model for Technological Change Applied to the Turbojet Revolution', *TC*, 14 (1973), 553–72.
R. Fries, 'British Response to the American System: The Case of the Small-Arms Industry After 1850', *TC*, 15 (1975), 377–403.
J.F. Hanieski, 'The Airplane as an Economic Variable: Aspects of Technological Change in Aeronautics, 1903–1955', *TC*, 14 (1973), 535–52.
D. Hounshell, 'Elisha Gray and the Telephone: On the Disadvantages of Being an Expert', *TC*, 16 (1975), 33–61.
T.P. Hughes, *Elmer Sperry: Inventor and Engineer* (Baltimore, 1971).
Louis Hunter, *Steamboats on the Western Rivers: An Economic and Technological History* (Cambridge, Mass., 1949).
R. Jenkins, *Images and Enterprise: Technology and the American Photographic Industry, 1839–1925* (Baltimore, 1976).
Arthur Johnson, *The Development of American Petroleum Pipelines: A Study in Private Enterprise and Public Policy* (Ithaca, N.Y., 1956).
David Landes, *The Unbound Prometheus: Technological Change and Industrial Development in Western Europe from 1750 to the Present* (Cambridge, 1972).
Otto Mayr, 'Adam Smith and the Concept of the Feedback System: Economic Thought and Technology in 18th-Century Britain', *Technology and Culture*, XII (1971), 1–22.
Elting Morison, *Men, Machines and Modern Times* (Cambridge, Mass., 1966).

David Noble, *America by Design: Science, Technology and the Rise of Corporate Capitalism* (New York, 1977).

Nathan Rosenberg, 'Technological Change in the Machine Tool Industry, 1848–1910', *Journal of Economic History*, XXIII (1963), 414–43, and idim (ed.), *The American System of Manufactures* (Edinburgh, 1969).

Thomas M. Smith, 'Project Whirlwind: An Unorthodox Development Project', *Technology and Culture*, XVII (1976), 447–64.

Langdon Winner, *Autonomous Technology*, (Cambridge, Mass. 1977).

4. Lynn White, Jr., *Medieval Technology and Social Change* (Oxford, 1962), p.78 Hereafter cited as White, *Medieval Technology*.

5. Lynn White, Jr., 'Technology in the Middle Ages', in *Technology in Western Civilization*, ed. by Melvin Kranzberg and Carroll Pursell, Jr., (New York, N.Y., 1967), vol.1, p.73. Hereafter cited as White, *Middle Ages*.

6. White, *Middle Ages*, I, 74.

7. Ibid., I, 73.

8. Ibid., I, 72.

9. Ibid., I, 72.

10. White, *Medieval Technology*, 62–5, 72–3.

11. Ibid., 59–60.

12. Ibid., 44, 78.

13. Karl Marx, *Capital: A Critique of Political Economy*, ed. by Frederick Engels (New York, N.Y., No Date), p.415. Hereafter cited as Marx, *Capital*.

14. Marx, *Capital*, 369.

15. Ibid., 370.

16. Andrew Ure, *The Philosophy of Manufactures* (London, 1835), p.20, quoted in Marx, *Capital*, 415.

17. Marx, *Capital*, 430.

18. Ibid., 457.

19. Ibid., 415–16.

20. Ibid., 416.

21. Ibid., 417.

22. Ibid., 458.

23. Ibid., 419.

24. Alfred D. Chandler, Jr., *The Visible Hand: The Managerial Revolution in American Business* (Cambridge, Mass., 1977), p.64. Hereafter cited as Chandler, *Visible Hand*.

25. Chandler, *Visible Hand*, 207.

26. Ibid., 208.

27. Ibid., 243.

28. Ibid., 241.

29. Ibid., 242–4.

30. Ibid., 324.

31. Ibid., 209.

32. Ibid., Chapter 10.

33. Ibid., 285.

34. Ibid., 411.

35. Ibid., 208.

36. Ibid., 240.

37. Ibid., 364.

38. Ibid., 244.

39. The section on 'power systems' is drawn from T.P. Hughes, 'Technology as a Force for Change in History: the Effort to Form a Unified Electric Power System in Weimar Germany', in *Industrielles System und Politische Entwicklung in der Weimarer Republik*, ed. by Hans Mommsen, D. Petzina, and B. Weisbrod (Düsseldorf, 1977), I, 153–66; 'Technology and Public Policy: the Failure of Giant Power', *Proceedings of the IEEE*, LXIV (September 1976); 'Managing Change: Regional Power Systems, 1910–1930', *Business and Economic History*, ed. Paul Uselding, Series 2, VI (1977), 52–68; and 'The Electrification of America: The Systems Builders', *Technology and Culture*, XX, (1979), 124–61.

40. Review of Kranzberg and Pursell, *Technology in Western Civilization*, by T.P. Hughes in *Isis*, LIX (1968), 207–8.

41. Theodore Roszak, 'Technocracy: Despotism of Beneficent Expertise', *The Nation* 209 (September 1969) reprinted in *Changing Attitudes Toward American Technology*, ed. T.P Hughes (New York, N.Y., 1975), 37–51.

Bronze Roman Piston Pumps

THORKILD SCHIØLER

The piston pump is one of the more advanced expressions of Graeco-Roman mechanical technology and yet it is a machine about which relatively little has been written. And notwithstanding the fact that twenty-one pumps have been recorded in modern times (see Table 1) much remains to be elucidated with respect to the machine's origins, methods of fabrication, purpose and, most obscure of all, the nature and the extent of influence of the legacy in pump technology.

The subject of ancient pumps is extremely interesting not least because they are fully and frequently described in the classical literature, first of all by Ctesibius in the third century B.C. and later on by Hero in about 60 A.D. These texts have been studied thoroughly by, amongst others, Dr Drachmann[1], and two of them, Vitruvius and Hero, are available in modern editions.[2]

With the manuscript sources there exists a variety of early illustrations of ancient pumps although none of these pictures dates from the classical period. There is also a large number of unpublished drawings of which Figures 2 and 3 are examples.

Section 27 of Hero's *Pneumatics* is devoted to the Fire Engine and Hero is specific in claiming that the device was 'used in conflagrations'. Vitruvius (Book X, Ch.7) is more general in referring to 'the machine of Ctesibius, which raises water to a height', and in claiming that 'water can be supplied for a fountain from a reservoir at a lower level'.

It is a commonly-held view that the Romans used piston pumps exclusively to fight fires. Bearing in mind the size and mode of construction of Roman pumps one is bound to envisage a very unequal contest in all but fairly minor blazes in houses, shops and the like. In fact the variety of locations at which the twenty-one archaeological specimens have been found strongly suggests that there were numerous applications of which fire-fighting may not have been the most important.

The Silchester pump[3] was found in a well which suggests water-supply, in this case from a depth of some 5 metres. The eight-cylinder pump from St Malo[4] was found in a water-tank on the sea-shore and whatever its purpose — draining a dry-dock perhaps, or filling salt-pans, or abstracting fresh water from some freak source — fire-fighting does not readily suggest itself. The Lake Nemi pump[5] — if the remains are genuinely those of a pump and the conclusion has been contested[6] — was found in the remains of two ships and is most plausibly to be associated with bilge pumping. Each of the four Cote d'Azur pumps, which will be considered in more detail later, were found in a wrecked ship. Perhaps they too were bilge pumps although

TABLE 1 The twenty-one Roman piston-pumps and their places of origin

Wooden pumps	Bronze pumps
Benfeld[8]	Bolsena A
Silchester[3]	Bolsena B
Metz[8]	Sotiel Coronado
Trier[8]	Civitavécchia
Trier[8]	Côte d'Azur
Trier[8]	Côte d'Azur
Zewen[8]	Côte d'Azur
Belginum[8]	Côte d'Azur
StMalo[9]	
Lake Nemi[5] p. 185	
Colle Mentuccia (Roma)[31b] p. 196	
Los Ullastres (Gerona, Spain)[10]	
Cap del Vol (Gerona, Spain) not published	

one cannot rule out the notion that they were part of the ship's cargo. Either explanation is significant.

The Sotiel Coronado pump, the principal subject of this paper, was found in a mine but its application is far from clear. Certainly drainage is as likely a use as fire-fighting. From a fairly late date comes the interesting but inconclusive evidence of Isidorus Hispalensis.[7]

A vessel which, when inflated from below, allows water to flow out, is called a syphon. Such vessels are used in the East. 'In the event of a fire in the house, they come running with syphons full of water and put the fire out, but they clean the vaults by forcing water up into the air.'

Roman pumps can be divided into two categories according to the materials from which they were made. Some were made of wood (plus lead in certain cases) and the others of bronze. A total of ten wooden

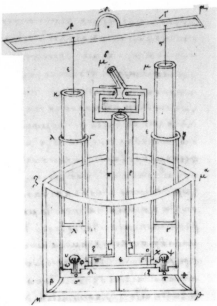

Figure 2. Manuscript figure to Hero's text. Facsimile of a previously unpublished manuscript figure. The Royal Danish Library, Thott 215, fol.14r.

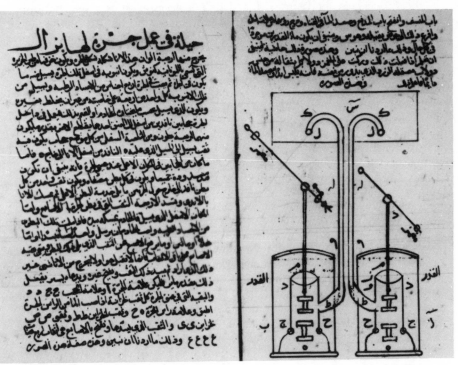

Figure 3. Philo of Byzantium. Manuscript figure in the Bodleian Library, Oxford, March 669.

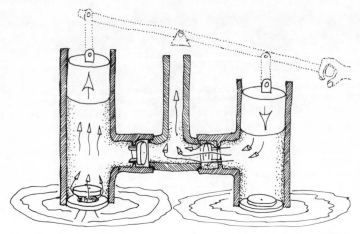

Figure 1. Flow diagram of double-piston pressure-pump.

pumps (if we include the Lake Nemi specimen) are known. They have been studied so thoroughly by Adolf Neyses,[8] R. Sanquer[9] and Frederico Foerster Laures[10] that it is not necessary to pursue the topic further here. Instead we shall concentrate on the bronze pumps and particularly on the question of how these intricate all-metal devices were made.

So far eight bronze pumps have been found. The earliest four finds are of the double-piston pressure-pump type (Figure 1) whilst the later four are of the single-piston pressure-pump type.[11] The two that are best known are in the British Museum. Less well known is a wonderful example — a very big pump discovered in 1795 — which has apparently disappeared in Rome. In principle the two sets of pumps, twin and single piston, are identical, but in the following we will look more closely at each example and supplement the material evidence with fragments from written sources.

THE MADRID PUMP (Figure 3A)

This well-preserved pump was found in 1889 in the copper mine of Sotiel Coronada, near the village of Valverde in the province of Huelva. The mine was at that time owned by the United Alkali Company Ltd of Liverpool. It had once belonged to Anselmo Boeck Meyer of Ceuta[12] and is first mentioned by R. Thouvenot, who assigned the pump to the first century A.D.[13] Today this technological rarity is exhibited at the Museo Arqueológico Nacional in Madrid. It is illustrated in Figure 4. The pump, which Gossé calls Spain's pearl

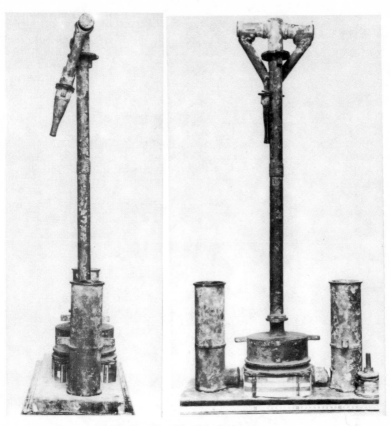

Figure 3A. The Madrid pump.

among Roman antique finds, is described in two papers from journals which are accompanied by a couple of sketches.[14]

EXAMINATION OF THE MADRID PUMP[15]

The pump consists of two cylinders with pistons. Each cylinder is connected to a common tank, which we will call a valve chamber because there are two valves in it. The riser pipe leading from the valve chamber has a height of almost one metre and bifurcates at the top into two branches, which meet again in the jet pipe (Figure 5). The area of flow is narrowed so much that the velocity of the water is increased twenty-five times. The jet pipe can be turned up and down and sideways, so that the jet of water can be directed as required, exactly as Hero describes.[16] The pump was cast by the lost-wax method and consists of twenty-six parts, only a few of which are missing including one of the valves and a few round discs.

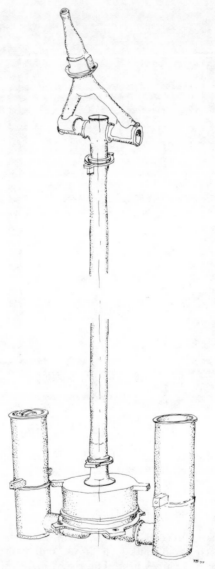

Figure 4. Sketch of the Madrid pump. Total height 1.35 m, width 0.41 m. Piston diameter 68 mm.

THE CYLINDERS (Figure 6)

The ratio between the cylinder diameter and the stroke of the piston is about 1:2, a value which is typical of Roman pumps. Let into the side of each cylinder near the base is a small diameter pipe with a flange for connection to the valve chamber. Each cylinder is reinforced with three

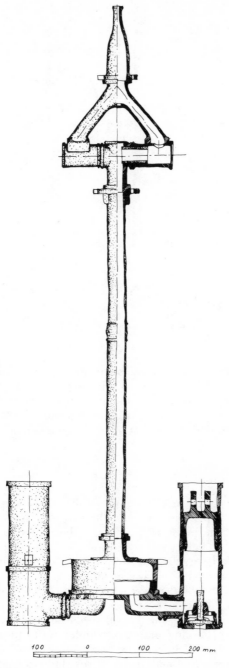

Figure 5. The Madrid pump. The design of this pump corresponds closely to Vitruvius's description.

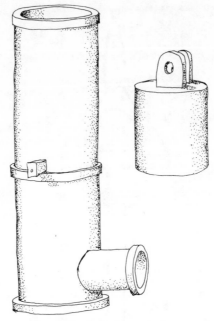

Figure 6. Cylinder and piston.

bands and adjacent to the central band are a pair of square lugs, one on each side of the cylinder body. Their purpose is not absolutely clear but very likely they were needed in the casting process. Such lugs are found on all bronze Roman pumps. A similar lug is visible on a bronze quiver exhibited in the Museo Nazionale di Villa Giulia.[17]

THE VALVE (Figure 7)

Each suction valve consists of three parts: the mushroom head with the valve spindle; the valve bushing; and the valve housing. The head and stem are cast in one piece, and although the part looks as though it were turned on a lathe, this was not the case. What have been presumed to be lathe tracks and centre-holes are more likely to be the result of the preparation of the wax model.[18]

In order to fill a cylinder on each rising stroke of the piston, water must pass between the face of the raised valve and its seat, then through the three openings in the valve housing, and finally upwards between the valve housing and the cylinder wall. Ideally these flow passages should be of identical cross-sectional areas and, so far as can be ascertained from the analysis, they are indeed comparable. It is evident that the Roman who made the Sotiel pump appreciated this

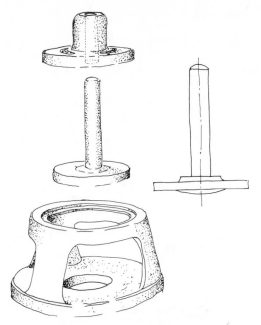

Figure 7. The lift valve construction consisting of: the mushroom head with the axial spindle; the valve bushing; and the valve housing.

important design feature. The whole mushroom valve structure constitutes a harmonious unit that bears comparison with a modern, well-designed, non-return valve.

THE PISTON (Figure 6)

The diameter of the piston is almost as great as its length and this helps to achieve a pressure-tight fit between the cylinder wall and the piston. Vitruvius (Book X, Ch.7) recommends that the piston 'be made smooth on a lathe'. Such a 'lathe' was probably a large, vertical fly-wheel turned by hand to which the bronze or brass work-piece is fastened for finishing and polishing. Most brass and bronze work-shops in India have such a 'lathe' but it is useless for turning.

Because of the casting technique used, the skirt of the piston is relatively thin and the space within the piston would have retained a volume of air when the pump was working, fortuitously providing a degree of absorbtion of pressure variations. A pair of identical lugs on the crown of the piston coupled with the piston rod by means of a 9 mm pin.

THE VALVE CHAMBER (Figures 8 and 8A)

This consists of two parts: a base of complex shape and a cylindrical upper casing. The base is fitted with two pipes leading into the valve chamber. The X-ray pictures show clearly that the valve chamber was made water-tight by turning a flange (turned in the wax, see notes 18 and 19) on the base and recessing it into the upper casing.

The top piece and base are connected by four pins shaped like the Greek letter gamma. This description of the shape comes from Hero.[20]

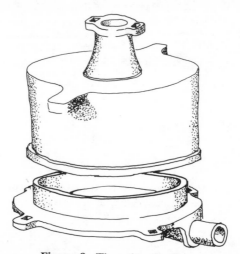

Figure 8. The valve chamber.

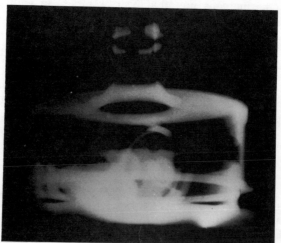

Figure 8A. X-ray picture of valve chamber with valves.

The gamma-shaped pins are provided with a transverse hole through which a small wedge is inserted (Figure 9). This same technique is referred to by Vitruvius (Book X, Ch.7) where he says 'Over the vessel a cowl is adjusted, like an inverted funnel, and fastened to the vessel by means of a wedge thrust through a staple...'. Moreover in the museum of Paestrum near Salerno in Italy there is exhibited a pair of compasses whose arms are connected by the same pin and wedge arrangement.

Vitruvius speaks also of the force acting on the upper casing of the valve chamber. If we assume that there is a pressure of ten metres of water inside the valve chamber, the force acting on the top piece of the Madrid pump must have been 180 kg or 1800 N. Hence each of the gamma pins had to carry a load of about 45 kg and their dimensions indicate that they were capable of this.

The upper casing of the valve chamber was cast in one piece from a wax model which was turned on a lathe perhaps of the type that Mutz[18] has reconstructed. On the crown of the upper casing is a flanged pipe connection. The flange has two square holes of the same size as the ten others in the various flange connections in the pump. As with the two cylinders, the upper half of the valve chamber has the same pair of lugs probably required in the casting process.

THE VALVES IN THE VALVE CHAMBER (Figure 8A)

X-ray pictures have shown that the delivery valves are still inside the pump, but it is not easy to form an entirely clear picture of their appearance. They are round and are fitted with four small orifices and possibly have a guide collar in the middle.

THE MAIN RISER (Figure 5)

It was a feature of the casting technique used that the fabrication of a long pipe was intrinsically difficult. The fact that shrinkage would cause a shortening of a few per cent was not a problem in itself. The overwhelming difficulty was that the mould prevented shrinkage altogether with the result that very high longitudinal residual stresses were induced. The risk therefore is that a pipe is at least weakened or might even snap; in this case it has failed just above the bottom flange.

The riser pipe has an outside diameter of 35 mm and is fitted near the base with a flange. The pipe was fitted into the neck of the valve chamber so that the two flanges were flush and could be clamped with a pair of gamma-shaped pins. Slightly above the middle of the pipe there is a band which does not appear to have any technical function and is presumably decorative.

At the top the pipe ends in a flange with two holes that are intended for the gamma pins. All the round and square holes in the pump were made in the wax model, and were not drilled or cut out in the finished bronze. The flange has a thickness of only 8 mm, which is not enough to guide the gamma pin, for which reason a fillet is provided, plane with the inner edge of the square hole. Hero gives details of the connection between the rising pipe and the nozzle assembly and suggests that the gamma pin should be soldered on.

THE NOZZLE ASSEMBLY (Figures 9 and 10)

The T-piece at the head of the rising pipe was cast in the shape of a cross after which the upper hole was blanked off by soldering on a circular cap. Fitted into the two horizontal outlets of the T-piece were short cylindrical tubes which allowed a swivelling connection with the two leading to the smaller diameter tubes that converge into the jet itself. These two short rotating pieces of pipe have reinforcing rings cast on each end and their outer extremities were closed by means of thin circular discs, soldered on. Oval holes in the pipes allowed the water to flow into the jet tubes. The connection was made by means of curved flanges cast on to the jet tubes and soldered to the pipes. It is a neat design which allows the jet to be rotated through a large arc. To prevent leaks a good fit between two cylindrical surfaces had to be achieved. Soldering the curved flanges of the jet tubes to the rotating pair of pipes must have been the last stage of assembly. At the apex of the converging jet tubes, and just under the jet itself, the casting was sealed with a circular disc soldered on after the casting core had been removed (Figure 10).

THE NOZZLE (Figure 11)

The cross-sectional area of the nozzle narrows so as to form a jet with a diameter of 8 mm. The nozzle is suitably designed from a hydrodynamic point of view.

The whole of this pipe structure is actually very complicated and could be designed much more elegantly nowadays. However, the interesting thing is that even in the most modern firefighting equipment we find the idea of the Y-shaped pipe whose origins can thus be traced right back to Roman times (see Figure 16).

THE TWO PUMPS IN THE BRITISH MUSEUM

The literature on these two pumps is rather extensive,[21] but the report on the excavation has never appeared in the English literature. From the Italian text it is clear that the pumps were found among bronze waste.[22] Six bronze objects are mentioned, such as '... a forearm holding a sphere; a little Hermes in bronze; a spearhead in bronze; a hare without a nose; an arm without a hand; a leg without a foot;

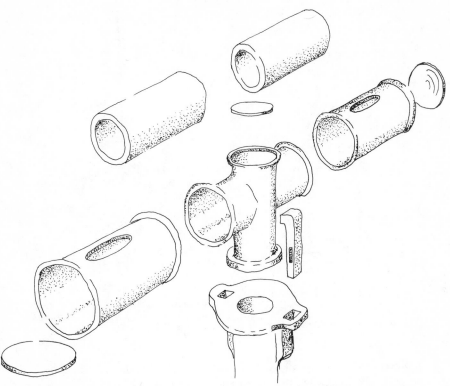

Figure 9. T-pipe and two pipes with oval holes. The gamma-shaped pin is a reconstruction.

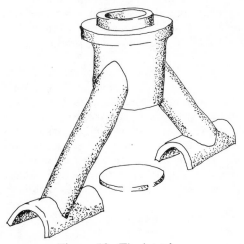

Figure 10. The jet tubes.

Figure 11. The nozzle.

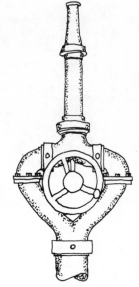

Figure 16. High-pressure nozzle used in modern fire-fighting equipment.

fragment of a statue of Aesculapius; coins from the Republican and Imperial Periods, the latest from Diocletian's time (A.D. 304)'.

There are also two components for the pumps but it has not been possible to identify them accurately. The possibility cannot be excluded that they have nothing to do with the pumps at all.

One of the few original contributions on the pumps is rarely mentioned in archaeological discussions because it appeared in a distinctly technical journal.[23] Magannis measured and drew pump A himself and his excellent drawing is reproduced in Figure 12. He emphasizes that:

> These valves consist of disks bolted or riveted loosely to the seats, thus allowing freedom to flap and open or close the passages as the rams were moved upwards or downwards. . . .
>
> Both the plungers are hollow and it has been suggested (note (21c), p.254) that they were formed in order to provide a means of introducing a plug of wood, perhaps covered with hide or leather, in order to make a watertight fit; but the author is inclined to believe that such was not the case (and so is the reporting author).

The reinforcing bands on pump B (Figure 13) have a very characteristic concave profile which we find on many Roman cocks.[24]

At the National Museum in Copenhagen there is an unpublished

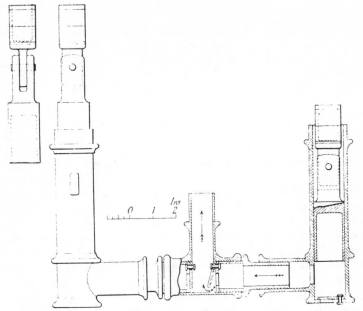

Figure 12. The best drawing of the British Museum pump A, made by P. Maginnis in 1911.

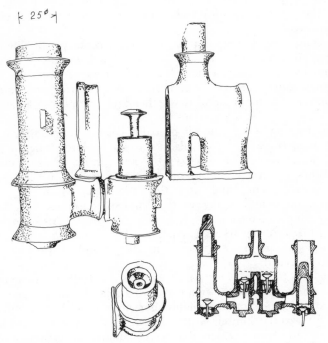

Figure 13. B.M.-pump B, with the guide ring for the mushroom valve spindle at the bottom.

cock (Inv. 8996) from Pompeii, which also has this concave band. In Pompeii there are more than twenty-five cocks of this type, and the Museo Nazionale di Napoli has more than fifty cocks with concave bands in store.

THE CIVITAVÉCCHIA PUMP, OR THE LOST PUMP OF ROME[25]

This pump — which has apparently disappeared — is only known through a drawing and a broadly descriptive article[26] (Figure 14). Since then a variety of new drawings have appeared. They are all very poor except one which is to be found in the Deutsches Archaeologisches Institut in Rome.[27] The original drawing is scaled in a very unconventional fashion: at the bottom there is a line of the same length as the diameter of the cylinder. Converted this means that the drawing is reproduced to a scale of 1:6. In other words the cylinder has a height of about one metre, or five times the height of the Madrid pump's cylinder.

This pump is very much larger than the others and capable of delivering twenty to thirty times as much water. In this case we may very well be looking at a fire-fighting engine (Figure 17). Visconti suggests that the rising pipe was made of lead and that its upper valve in some way supplemented the two below. It is difficult to accept that it could have served any such purpose and more likely it was used to isolate the pump mechanism from an elevated water tank if ever repairs were necessary.

Visconti assigns the pump to the second century A.D. The grounds for this dating are rather slender, namely that the pump is a fine piece of handicraft and that the aqueduct in Castronovo bears the name of Antoninus Pius. There is a lead pipe stamped with this name from Castronovo and the question is whether this pipe once belonged to the pump.[28]

The drawing of the pump is probably correct in principle, but there are several errors in the details. It would be interesting to find the pump[29] which is mentioned by H. B. Walters in his bronze catalogue.[30]

Walters compares the pump in the British Museum with the Civitavécchia pump, and says that the latter is in the Vatican. However, the pump is not in the Vatican museums. On the other hand, a couple of references[31] lead to Il Magazzino Archeologico, also called Museo Antiquarium Comunale. This museum was opened in 1894, reorganized in 1906, closed in 1949,[32] and in 1979 plans were made to reopen it.

The first, second and third editions of Helbig's famous guide contain no mention of the pump.[33] Two Italian engineers have studied water and engineering thoroughly but neither of them mentions the pump. Squassi has himself visited the Museo Antiquarium and includes photos of three water heaters that were exhibited in the museum.[34] Two exhibitions have been held in Rome (1910 and 1938) at which objects

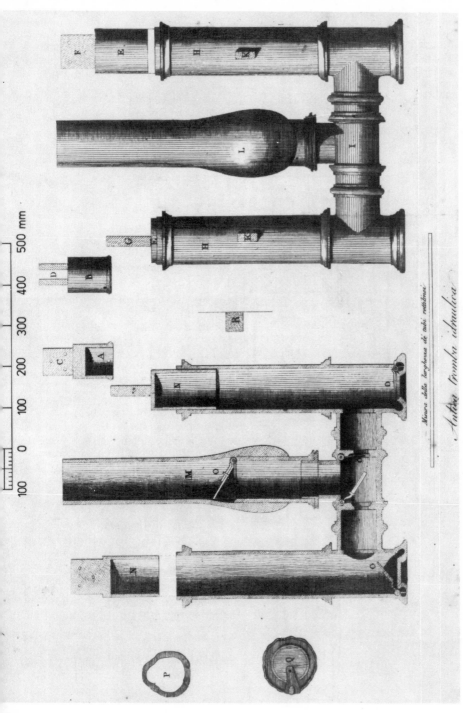

Figure 14. The Civitavécchia pump. Facsimile of the drawing in Ref. 26, *Opere Varia II.*

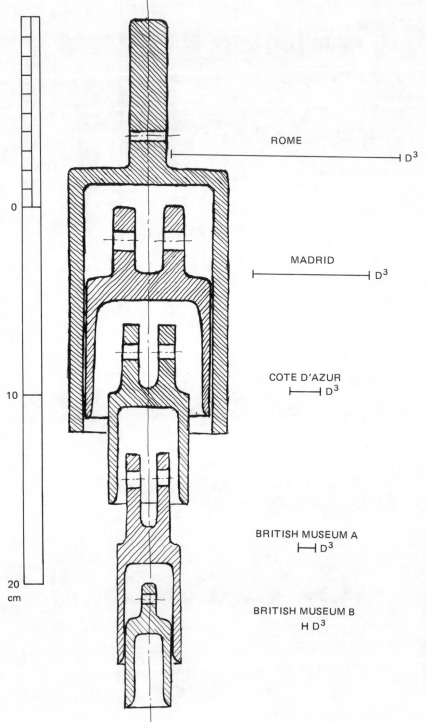

ROME
⊢D³

MADRID
⊢————————⊣D³

COTE D'AZUR
⊢——⊣D³

BRITISH MUSEUM A
⊢—⊣D³

BRITISH MUSEUM B
H D³

Figure 17. Comparative sketch of pump sizes showing to scale the relationship of piston diameter to diameter cubed.

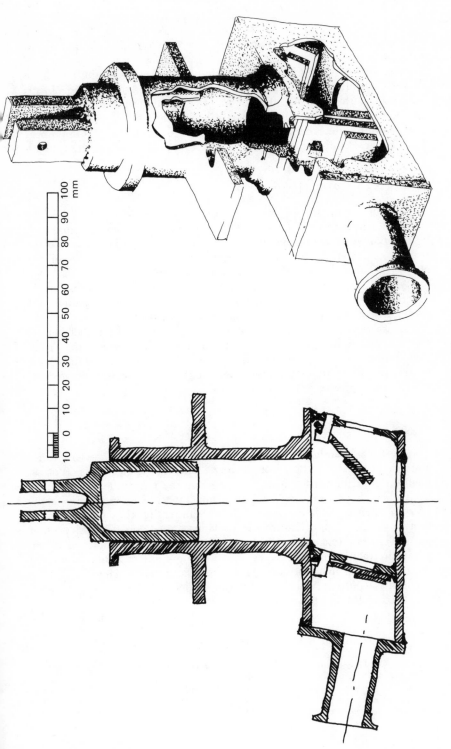

Figure 15. Two views of the Côte d'Azur pump.

connected with hydraulic engineering were displayed but the pump did not appear in either.[35]

The Museo Antiquarium Comunale now comes under the Direzione Musei Capitolini and here there is a hand-written list of the objects contained in the now closed museum. The pump is not included in the list. Some of the old museum objects are in five hundred crates that are now standing in Palazzo dell'Esposizione. Conceivably the pump is 'lost' in one of these crates.

THE CÔTE D'AZUR PUMPS

These four, identical pumps were found 20 km west of Cannes in the wreck of a Roman ship. Georges Rouanet's paper on these pumps, with its ninety-five drawings and photos, is a model exposition to which the reader is referred.[36] The pumps can be dated to within ten years of A.D. 50 on the basis of other objects found in the ship.

As will be seen from the drawing in Figure 15 these pumps have a different arrangement from the four mentioned earlier although the principles of the individual components are largely the same. The piston is hollow and fitted on its crown with a pair of parallel lugs to connect the piston rod. According to the text and drawings these lugs were bronze-welded to the piston[37] but that is not the case. The square dowels on the cylinders and the lugs on the crown of the piston were cast in one with the rest of the material.

CONCLUSION

The 'lost-wax' method of casting has been known since ancient times and is still used today; complicated machine components (not to mention gold fillings for teeth) are cast by the technique. Quite apart from their importance as hydraulic machines, Roman pumps show every evidence of being an ancient example of the application of 'lost-wax' casting and contrary to previous notions on the subject their parts were not turned on the lathe. Nowadays we can make a piston pump in many different ways, but the only method that does not require rotating machine tools is the lost-wax method. Here, all that is needed is a potter's wheel for the wax model. That is apparently what was used in Roman times.

Notes

1. A.G. Drachmann, *The Mechanical Technology of Greek and Roman Antiquity*, Copenhagen, 1963, p.155.

2. *The Pneumatics of Hero of Alexandria*, a facsimile of the 1851 Woodcroft Edition, introduced by Marie Boas Hall, London and New York, 1971. Vitruvius, *The Ten Books on Architecture*, translated by Morris Hicky Morgan, New York, 1960.

3. G.C. Boon, *Roman Silchester*, London, 1957, pp.159–161.

4. See J.G. Landels, *Engineering in the Ancient World*, London, 1978, p.83.

5. G. Ucelli, *Le Nave di Nemi*, Rome, 1950, p.196.

6. See, for example, S. Shapiro, 'The Origin of the Suction Pump' in *Technology and Culture*, Vol.V, No.4, pp.566–574.

7. Isidore of Seville, *The Etymologies (Origines)*, Book XX, Chapters 6 and 9.

8. Adolf Neyses, 'Eine römische Doppelkolben-Druckpumpe aus dem Vicus Belginum', *Trierer Zeitschift*, Vol.35 (1972), pp.109–121. An abbreviated version of the same article is to be found in *Technikgeschichte*, Vol.39, pp.177–185.

9. R. Sanquer, in *Gallia*, Tome 31, 1973.

10. Frederico Foerster Laures, 'Los Ullastres; Discovery of Objects which may be a bilge pump, in the wreck of the first century AD ship' in *International Journal of Nautical Archaeology*, Vol.8, No.2, pp.172–4, London, 1979. See also *The Pneumatics of Hero of Alexandria*, op. cit., p.45.

11. The last four, identical pumps were found in 1970 in a shipwreck near Cannes. They have been published by Georges Rouanet, 'Etude de quatre pompes a eau romaines', *Cahiers d'archéologie subaquatique*, numéro III (1974), pp.49–79.

12. Some years after the find, Anselmo Boeck Meyer offered the pump to The British Museum for 121 guineas. Then in about 1930, he offered it to the museum in Leeds for £500, but the museum was only prepared to pay £150. Just before the Spanish Civil War, the Spanish Government bought the pump for 20,000 pesetas. This information is taken from a document in the Museo Arqueológico's archives, privately communicated by José A. Garcia-Diego of Madrid.

13. R. Thouvenot, *Essai sur la province romaine de Bétique*, Bibliothèque des écoles francaise d'Athenes et de Rome. Paris, 1940, p.257 and Figure 16.

14. G. Gossé, 'Las minas y el arte minero de España en la antigüedad', *Ampurias*, Vol.IV (Barcelona, 1942), p.57. J.M. Luzón, 'Los sistemas de desagüe en minas romanas del suroeste peninsular', *Archivo español de arqueologia*, Vol.41 (Madrid, 1968), pp.117–120.

15. In April 1974 I had the opportunity of examining and measuring the pump. This research was financed by the Danish Research Council for the Humanities, and contact with Museo Arqueológico Nacional was arranged by the Deutsches Archeologische Institut in Madrid. Since then, José A. Garcia-Diego, with the valuable assistance of the firm of Laboratorio Geocisa, Madrid, has had the various components of the pump X-rayed and gamma-rayed. In the X-ray examination a voltage of 150–165 kV was used at a distance of 70–700 mm. In the gamma-radiation test, an intensity of 60 Ci was used, with ^{192}Ir, at a distance of 900 mm. Copper and lead filters and absorption plates were used as protection against the external radiation.

16. *The Pneumatics of Hero of Alexandria*, op. cit., p.45.

17. Alessandro della Seta, *Museo di Villa Giulia*, Rome, 1918, p.225, Inv. No. 6894.

18. That it was the wax model which was turned is not the view of Alfred Mutz, *Die Kunst des Metalldrehens bei dem Römeren*, Basel & Stuttgart, 1972, but it is, on the other hand, shared by J.F. Cave, 'A Note on Roman Metal Turning', *History of Technology*, Vol.2, (London, 1977), pp.84–85.

19. Alfred Mutz, op. cit., and J.F. Cave, op. cit. This production method is well known in India and the Far East. Rustann J. Mehta, *Handicrafts and Industrial Arts of India*. Bombay, 1960, p.34. '. . . a layer of prepared wax, fitted in a lathe and turned. . .'.

20. *The Pneumatics of Hero of Alexandria*, op. cit., p.45.

21. (a) *Jahrbuch des Kaiserlich Deutschen Archäologischen Instituts*, Vol.VIII, supplement: 'Archäologischer Anzeiger, p.186.

(b) Anon., 'Old Roman Pumps', *The Engineer*, Vol.78 (1894), p.4.

(c) Frederick Davies, 'Notes on a Roman Force-Pump Found at Bolsena, Italy'. *Archaeologia*; or 'Miscellaneous Tracts Relating to Antiquity', Vol.LV (London, 1897), pp.254–256.

(d) H.B. Walters, *Catalogue of the Bronzes, Greek, Roman and Etruscan, in the Department of Greek and Roman Antiquities*, British Museum, London, 1899, pp.331–332.

(e) *A guide to Exhibition Illustrating Greek and Roman Life*, London 1908, first ed., p.110.

(f) Science Museum, *Handbook of the Collections illustrating pumping Machinery*, by G.F.

Westcott, London, 1933. Part II, Descriptive Catalogue, p.28, and Plate II facing p.19.

22. *Notizie degli scavi di antichitàs*, 1891, p.121.

23. James P. Maginnis, 'Notes on Bronze Pumps in The British Museum', *Institution of Mechanical Engineers, Proceedings Year 1911*, pp.399–401 and 402, and Plate 21.

24. (a) Fritz Kretzschmer, 'Römische Wasserhähne, *Jahrbuch der Schweizerischen Gesellschaft für Urgeschichte*, Vol.48, pp.50–62.

(b) Enzo Fassitelli, *Duemila anni di impianti in Italia; Tubi e valvole dell'antica Roma*, ed. Petrolieri d'Italia, Milan, 1972.

25. The same pump is also known as the Castronovo pump, after the antique town of Castrum Novum. Two hundred years ago this was called Chiaruccia, but today it carries the name S. Marinella.

26. Ennio Quirino Visconti, 'Discrizione di un'antica tromba', *Giornale della letteratùra italiana*, Vol.V (Mantova, 1795), pp.303–307. (Only a few copies of Vol.V are to be found, in Italy.) The article was reprinted with a new drawing in E.Q. Visconti, *Opere Varia II*, Rome, 1829, pp.29–32.

27. See in catalogue under E.Q. Visconti. The drawing is a blueprint, signed in ink by General Carl Giebeler, Grosslichterfelde (near Berlin) on 5 December 1911.

28. *Corpus Inscriptionum Latinarum*, Vol.XI, pars 1, No.3586, Berlin, 1888.

29. Dr John Peter Oleson, American Academy in Rome, also searched for the pump in 1977, independently of me. See also J.P. Oleson, 'Research on Greek and Roman Pumping Technology' in *Current Anthropology*, Vol.20, No.3, September 1979, pp.655–656.

30. H.B. Walters, op. cit., pp. 331–332.

31. (a) Guido Ucelli, *Le navi di Nemi*, second ed. Rome, 1940, pp.182 and 195–197.

(b) Ettore de Magistris, *La Militia Vigilum della Roma Imperiale*, Rome, 1898, p.92.

32. The opening of the museum is mentioned in *Bullettino della Commissione archeologica comunale di Roma*, Serie 4, Vol.22, pp.145–146. 'La sala VI . . . una pompa aspirante e premente.'

33. W. Helbig, *Führer durch die öffentlichen Sammlungen klassischer Altertümer in Rom*, third ed., 1912, p.577.

34. Francesco Squassi, *'L'arte idro-sanitaria degli antichi'*, *Epoca preromana e romana*, Tipografia Filelfo, Tolentino, 1954. Francesco Pellati, 'L'ingegneria idraulica ai tempi del l'impero Romano', *La scienza e la tecnica ai tempi de Roma imperiale*, Vol.XII, Roma, 1940.

35. *Catalogo della Mostra Archaeologica nelle Terme di Diocleziano*, Esposizione internazionale di Roma, Bergamo, 1911. *Mostra Augustea della Romanitá*. Catalogo, third ed. Without town and year (Rome, 1938).

36. Georges Rouanet, op. cit.

37. In December 1978 I had the opportunity of seeing the pumps together with Dr Georges Rouanet in the little archaeological museum in Frejus. All four pumps have been cleaned and restored. Some of the valves are more than just watertight — they are airtight. As far as I could see, there were no bronze welds — everything was cast in one piece. The inscriptions and markings were made in the wax, not in the bronze.

Measurement in Galileo's Science*

STILLMAN DRAKE

I

Measurement is so frequently used in ordinary affairs that units of length, time, angle and weight form part of everyday language. Ancient languages show this to have been always the case, and many remains of antiquity in the form of weapons, ornaments, sculptures and edifices show that accurate measurements were common in early technology. In early science, however, careful measurement appears to have been confined to astronomy and optics. In physics, which began with Aristotle as the science concerned with motion, actual measurement hardly entered before the time of Galileo.

Were it not for astronomy and optics, historians of science would probably always have said that the role of actual measurement in natural philosophy was negligible until the Scientific Revolution, and then became so prominent as to forge links between science and technology which transformed European society. In that case the importance of measurement in Galileo's science could not have escaped attention, and the history of measurement might have received some of the deep study that in recent decades has been lavished on the history of philosophical conceptions.

As it is, the little that I can find about the history of measurement deals mainly with ancient units, a subject that (perhaps significantly) fascinated sixteenth-century humanists but remains largely of antiquarian interest. Historians of technology have indeed recognized the ingenuity of builders and craftsmen in assuring the accuracy of direct measurements, and of surveyors and tunnellers in making indirect measurements where direct measurements were impossible. But little appears to have been done to explain why early physics ignored actual measurement, or how that ultimately came to transform the very basis of science. Hence I shall begin by hazarding some answers to those questions.

---

*Galileo has long been considered the first philosopher to investigate the theoretical reasons underlying observed practical effects of interest to craftsmen; others before him had, of course, attempted to improve technical practice — not at all the same thing. I invited Professor Drake to consider Galileo in relation to measurement, an activity of importance to both science and craftsmanship. (A.R.H.)

It certainly seems that physics as the science of motion could not neglect actual measurement or, at least, that concern with measurement in astronomy as the science of celestial motions would necessarily spread quickly into physics as the more general science. But that ignores a semantic shift in the word 'science'. The kind of systematic astronomy we think of came into being only a century or more after Aristotle had defined the goal and method of science. Whether his definitions, retained for many centuries, would have included that kind of astronomy is something that must be determined historically, not by current criteria of science.

For Aristotle, science was the understanding of things in terms of their causes, which are not given directly by experience but must be supplied by reasoning. Scientific knowledge was accordingly distinguished from knowledge attained through practice and experience, which Aristotle called *techne*. He did not disparage practical knowledge, but set it apart from scientific knowledge as different in kind, not comparable with and hence basically irrelevant to it. It is therefore not surprising that when Aristotle wrote his *Physics*, he did not concern himself with actual measurements of motion even in discussing speeds, times and distances, let alone forces and resistances. Actual measurements had nothing to do with causal analysis and could not alter the science of physics.

A parallel treatment of cosmology was given in Aristotle's book *On the Heavens*. If we think of its system of homocentric spheres as an astronomy, it was an astronomy without positions, angles, distances or speeds of the heavenly bodies. Cosmology was a schematic description into which knowledge attained by practice and experience could be fitted. Eudoxus had shown that uniform rotations of spheres with their axes suitably oriented in other spheres would appear from a central fixed earth to give motions of the kinds observed in the heavens, some uniform and some non-uniform and even changing direction. Aristotle supplied the causal driving-apparatus needed for the science of cosmology. Anything else would be *techne* — and, when we look at Hellenic astronomy as *techne*, many seemingly awkward historical problems simply vanish.

During the Hellenic period, from 300 B.C. to 300 A.D., a great deal of interest and no little skill were shown in actual measurements ranging from celestial motions to the design of catapults. The semantic shift already mentioned induces us to regard that activity as representing advancing science, though it did not advance, retard, modify or in any way alter the causal reasoning of Aristotelian natural philosophy. The case of Hellenic astronomy is particularly interesting, since we know a good deal about the origin of systematic mathematical astronomy and its reception by natural philosophers.

When Hipparchus came to use historical and contemporary data obtained by careful measurement of times and angles, to predict eclipses from assumed uniform rotations, he found it necessary to introduce eccentric motions.[1] These could have been viewed as contradicting Aristotle's cosmology, but historically they were not. Mathematical astronomy was welcomed for its interest and utility, scientific cosmology for some other reason, perhaps elegance or religious significance. At any rate Geminus, a contemporary of Hipparchus, simply forbade astronomers to consider causes of celestial motions, which they neither were trained to do nor needed for their work.[2] All causal explanations were left to physicists, who were natural philosophers. Centuries later, that still satisfied Simplicius who reported no alterations in Aristotelian cosmology as having been made or as necessary. Ptolemy, who codified Hellenic astronomy, explicitly excluded both physics and metaphysics from his *Almagest* to obviate philosophical disagreements, as he said, and his book survived in peaceful co-existence with *On the Heavens* for a very long time. In effect, astronomy was treated as *techne* rather than as science in both the Arabic and Latin cultures of the Middle Ages.[3]

On that basis I may say that the history of measurement from the time of Aristotle to that of Copernicus, Tycho Brahe, Kepler and Galileo belonged entirely to technology, astronomical and optical measurement included. After the Aristotelian cosmology had been challenged in a way that did not admit of compromise, Galileo added a similar challenge to Aristotelian physics. New kinds of astronomical measurements, as well as more precise ones, entered into the challenge.[4] Kepler offered a completely new basis for cosmology, while Galileo rejected any part of that science that was not confirmed by accurate measurements. At the same time he introduced actual measurements into physics, and, as someone has said, began to measure what could be measured but had not been, and to make measurable things that had not been measurable.[5]

II

Everyone will admit that speeds of motion can be measured, and most will grant that they had not been actually measured before Galileo. Some may deny that Galileo made measurable anything that had not been *measurable*, supposing me to have meant only things that had not been *measured*. Yet I agree with the unknown author, believing that the objection rests on insufficient attention to the nature of measurement itself and to the fact that its historical development reflects internal necessities and not just accidents.[6] This is a very complex subject that has not been given the attention it deserves by historians of science. In this paper I can deal with it only by illustrative examples relating to Galileo.

Consider the case of surface tension. It began to be measured when it was given a name, in the eighteenth century. But Galileo made it measurable, though he did not measure it. Someone might contend that

Galileo's Geometric and Military Compass (*c.* 1605)

that is nonsense; anything now measurable was always measurable, even though a means of measuring it may have been lacking. I myself once shared the utter confidence one feels in such tautological statements, until serious historical researches taught me the greater value of trying to get the hang of actual events. Logic cannot be violated, but in understanding either history or science there are also questions of meaning and interpretation.

Galileo noted that a small resistance to sinking seems to exist at the surface of water and not within it, but he was reluctant to assign different properties to water according to its position. In 1611 he studied the relative depths to which floating chips of different materials denser than water descended below the surrounding surface. Those depths were observably related to the densities; invoking the principle of Archimedes, Galileo declared them to be directly proportional. That made surface tension measurable, though Galileo neither named it nor measured it, and when it began later to be measured, the procedures used were different.

Now, I should consider it merely misleading to say that surface tension was measurable by Archimedes, who did not even notice the phenomenon. It would be even more misleading to say that Archimedes, or for that matter Galileo, failed to measure surface tension, because in the normal way of talking (except among historians of science) we do not say that people have failed to do something they have not tried to do. Galileo analysed something, and published his results, which made surface tension measurable by anyone to whom that concept (repugnant to Galileo) happened to occur. It did not occur to anyone for a long time, and then mainly in connection with capillary action. Once surface tension was given a local habitation and a name, it was seen to be measurable, was measured, and for us now explains the floating of dense laminae and the standing of drops of water on cabbage leaves, both of which were noticed by Galileo, linked together, and explained as a resistance by dry surfaces to wetting. It takes a bit of force, he said, to spread a drop of water over dry wood, or to separate from a surface the fluid wetting it. What is implied by an observed phenomenon is a matter of logic, but not only of logic. Awareness of other phenomena, focus of attention on what is to be explained, and preference for one kind of explanation over another are examples of non-logical factors in seeing what is implied by observation.

To observe relative depths of floating, estimating them by eye to be proportional to densities, may be regarded as measurement or may be called a hypothesis testable by measurement. When such a hypothesis is not arbitrary, but applies a principle previously confirmed by many and varied measurements, it is quite usual not to bother with difficult procedures but to subsume the observations under the principle. Sometimes errors arise in that way; more often, the hypothesis affords better indirect measurements than could be obtained directly. From Galileo's time on, scientific measurement has been a mixture of actual and hypothetical procedures of this kind. Before actual measurement entered science, things were more clear-cut; *techne* advanced by means of it, while natural philosophy remained untouched by it.

What I have just called hypothetical procedures in measurement has value to science because these procedures are tied to other actual measurements and it is known how to make direct measurements to test them. A different kind of hypothetical measurement was introduced into natural philosophy in the Middle Ages, mainly in the fourteenth century, which I shall call theoretical measurement. This was of great value in the classification of motions as uniform or non-uniform with respect to time, or with respect to the parts of the moving thing.[7] Theoretical measurement employs units, or measures, that generally differ from those capable of use in actual measurement; medieval examples are 'one degree of coldness' and 'the speed at the middle instant'. Like the cosmology of Aristotle, the science of motion thus created was schematic and could neither be assisted nor refuted by actual measurements. Parts of the medieval science of motion were later found to coincide with actual measurements carried out by Galileo, inducing historians to believe that he adopted those parts on faith. Other parts did not coincide with any actual measurements ever made, remaining of antiquarian interest in the history of science even if fruitful in mathematics.[8]

Theoretical measurement was less like actual measurement than it was like another technique of medieval natural philosophy, called 'proceeding in imagination', which corresponded to what are now called 'thought-experiments'. Unlike actual experiments, these are intended not to obtain new data concerning natural phenomena, but to test the language in which we describe them in order to see if it leads to paradoxes or contradictions when extended beyond the range of practicable procedures. Thought-experiments were a principal medieval contribution to science, which has never ceased to make extensive use of them.

Theoretical measurement and thought-experiments dominated natural philosophy until Galileo introduced experimental measurements. These seek new data from nature within the range of practicable procedures, concerning things of kinds not previously measured. Optics provides some earlier examples, and perhaps experimental measurements of specific gravities preceded the hydrostatics of Archimedes. *Techne* generally, however, was mainly applied mechanics, and once the lever law was known there would seem to have been little occasion for experimental measurements of the kind about to be discussed. In any event none had been made of speeds in accelerated motion, which became the foundation of Galileo's science.[9]

Until fairly recently it remained a matter of conjecture how Galileo had gone about the determination of speeds in natural descent. That has now become a matter of reconstruction from some of his unpublished papers providing evidence of the introduction of actual measurement as a basis of the mathematical theorems contained in his final book.[10] That move had been preceded by the traditional causal approach of natural philosophy, vestiges of which impeded Galileo later in correctly interpreting the first measurements he made and kept him from finding at once a sound derivation of the law of fall disclosed by them. His shift from the science defined by Aristotle to one resembling modern science resulted from experimental measurements and their interpretation. It was accordingly more like an extension of *techne* than of natural philosophy, though most who came after him laboured to restore the latter in some form.[11]

III

Galileo's first reasoning about speeds in natural motion is found in his *De motu* ('On motion'), composed at Pisa in 1591–2 with the intention of publication. His arguments were causal, taking weight as the cause of downward motion but rejecting the absolute weights supposed by Aristotle. Nothing in *De motu* was derived from actual measurement; theoretical measurement and thought-experiments predominated in its arguments. Galileo supposed that acceleration takes place only briefly in fall from rest, on causal grounds outlined in antiquity by Hipparchus. It was therefore ignored in Galileo's first account of speeds along inclined planes, because he at first supposed that a different uniform speed is associated with each slope, and that for a given weight the speed is determined by the component of weight in the vertical. This also was argued causally and led to some illusory conclusions about ratios of speeds. Galileo admitted that these had not been borne out by actual test, explaining that friction, air resistance, imperfectly hard materials, and irregularities of surfaces constituted irremovable 'material impediments'.[12] Like the causal approach, Galileo's explanation of disagreement between mathematics and experience was Aristotelian in spirit. In challenging some of Aristotle's conclusions at this time, he did not attack his conception of science.

Galileo's next ten years, at Padua, were devoted mainly to *techne*. He wrote treatises on mechanics, military architecture, and the measurement of distances by sighting and triangulation. An instrument he devised for that purpose was made useful also for measuring and calculating in general. Thus he had come to know a good deal about actual measurement when, in 1602, he returned to the study of natural motion. In that year he discovered his first correct and fruitful theorem about motions on inclined planes, still based on the incorrect assumptions of *De motu*. From this he conjectured that the time of descent should be the same along any circular arc to its lowest point, but was told by Guidobaldo del Monte that his own experiments did not confirm this. Guidobaldo had used a ball rolled in the rim of a vertical hoop; Galileo recommended observation of long pendulums to reduce material impediments and procedural difficulties. In a postscript he added:

> When I have a little leisure I shall write . . . of an experiment that has occurred to me for measuring the force of percussion. . . . What you say . . . is well put, and when we commence to deal with matter, then by reason of its accidental properties the propositions abstractly considered in geometry begin to be altered, from which, thus perturbed, no certain science can be assigned — though the mathematician is so absolute about them in theory.[13]

Clearly, Galileo was not only engaging in experiments but had begun seeking ways to measure things previously unmeasurable. Yet he had not changed his view of the inapplicability of mathematics to material science. With regard to motion, at least, the accidental properties of matter stood in the way.[14]

Probably experiments with long pendulums at this time directed Galileo's attention to the necessity of taking acceleration into account in natural motion from rest. During 1603 he solved some interesting problems independently of this by applying his 1602 theorem, but some of his notes show that he realised he would not advance much further without knowing how speeds changed during acceleration. He tried two conjectures, one based on medieval impetus theory and the other on a rule, also formulated in the Middle Ages, that in uniform acceleration from rest, the terminal speed is double the overall speed from first to last. Neither conjecture gave him any useful clue to change of speed during acceleration. Early in 1604 it occurred to him to try to measure experimentally some speeds along an actual inclined plane, much as it had occurred to him in 1602 to try to measure the force of percussion, though that attempt had not been productive of useful results.[15]

Galileo had previously tested conclusions which he had reached by reasoning, sometimes without success (as in the case of his *De motu* rules) and sometimes with good agreement (as in the case of his 1602 theorem). Merely testing a conclusion focuses the attention on a success-or-failure result. Seeking a measure for its own sake, on the other hand, raises broad questions of meaning and procedure. Now, up to this time 'speed' had never been defined, though Aristotle had defined 'equal speed' and 'greater speed' in terms of distances traversed and times elapsed.[16] That meant that the theory of proportion was applicable to speeds, whatever they might be. Distances could be measured as accurately as one wished, while brief times could not be accurately measured. That did not matter if equal times could be assured, since speeds during equal times would necessarily be proportional to the distances traversed, and those could be measured. Brief times can be equalized by musical beats, whence it was possible to measure relative speeds.[17]

Using a plane of very gentle slope, and beats of about half a second each, Galileo measured the distances traversed from rest by a ball during each of eight successive equal times. His length-units were *punti* of about 0.95 mm, and his rhythym-judgement was accurate within 1/50 second. Consequently the numbers he recorded gave him, almost at once, the rule that distances from rest are in the same ratio as the squares of the respective times. The measured distances agreed with calculation within 2%, which meant that a science of natural motion no less in agreement with experience than the mechanics based on the lever law was perfectly possible. Realization of this doubtless delighted Galileo as much as it astonished him, considering that only a year before, when discussing experiments of motion along inclined planes and in circular arcs, he still doubted the possibility of any 'certain science' of such matters. For under the erroneous causal assumptions of *De motu*, discrepancies with experiment had been very great so far as speeds in general were concerned.

The times-squared law enabled Galileo to relate distances and times correctly in a great variety of theorems and problems, which he proceeded to write out during the next few years. Those quantities were directly measurable. But Galileo's procedural reasoning for measurement of speeds had merely misled him. To follow out his initial misapprehension and its eventual correction is highly instructive regarding the processes of scientific measurement. Had Galileo's science of motion been no more than a kind of *techne*, it would have stopped with a firm rule relating two kinds of measurable quantities — distances and times. That would suffice for anything of practical utility, as it sufficed for Galileo's theorems. But what Galileo had set out to measure were the changing speeds during descent from rest, which appeared to him to have been made measurable uniquely, if indirectly, and which he wished to incorporate into his science of motion. In that regard he was misled by the reasoning behind his procedure, for that made speeds proportional to distances from rest.

Three years went by before Galileo realized that speeds in accelerated motion are proportional not to distances but to square roots of distances. That alternative had occurred to him in 1604, but he had decided against it.[18] Certainly it was not a very plausible assumption from which to derive the times-squared law, of which he was quite certain by actual measurements. In October 1604, after mulling over possible ways to derive the law without finding any certain and indubitable basis, he decided on an assumption that was 'physical and contained much of the evident'.[19] This was that speeds are as the distances from rest regardless of times — a generalization from his original reasoning that was incorrect, but that Galileo used at this period for a supposed derivation of the law of fall. Without knowing what had gone before it, historians of science offered various explanations for Galileo's defective assumption, including even the preposterous notion that he had misapplied the medieval middle-speed postulate which, if correctly applied, would have given him the rigorous derivation he sought.

Yet Galileo had explicitly said, in writing out his supposed derivation, that his assumption was physical (not theoretical) and was borne out by machines which act by striking. Proportionality of speeds to distances from rest does indeed seem to be borne out by such devices, which do in fact deliver blows proportional to the distances through which a given weight has fallen. Galileo reasoned, naturally enough, that since the weight remains the same, only its speed changes, whence it must be proportional to the distance fallen. This confirmed physically the reasoning used in making the original measurements, so he rejected the alternative of square roots of distances. His wrong choice at this stage was reinforced by his knowledge that square roots of distances were proportional to times, which he had always regarded as the *inverse* of speeds in motion.

This shows how difficult it may be to make sure just what it is that has been measured experimentally. Equalizing of times had been adopted by Galileo as a necessary device for measuring speeds at all, after it directed his attention to distances measured — but to distances taken discretely, as if time grew by units and not continuously. It took Galileo quite a while to realize that during acceleration, equal times give equal additions of speed, and he found that out only circuitously, by going back to square roots of distances. The explanation of his erroneous assumption in October 1604 was not poor reasoning but bad luck — his misfortune in hitting upon a physical phenomenon — the force of percussion, which he had previously investigated — that seemed to depend only on speed, whereas it depends on energy measured by the square of speed.

It may be asked why Galileo returned to the square root of speed after having once rejected it. The answer here has to do not with experiment but with mathematics. In a later attempt to establish another proposition, relating to speeds of descent along different inclines at equal vertical heights, Galileo encountered a paradox or contradiction with his 1604 derivation. That was resolved, late in 1607, by moving from discrete to continuous change of speed, mathematically admitting square roots of distances. These in turn promptly suggested another set of experimental measurements to confirm Galileo's old conclusion in *De motu* that horizontal motion would be uniform in the absence of material impediments. A by-product of these measurements was the discovery of the parabolic trajectory, which was then followed by the recognition that speeds in acceleration were proportional directly to times, effectively completing Galileo's science of motion.[20]

That Galileo's confidence in his new science was grounded on its agreement with careful measurements is beyond doubt, for he had no rigorous demonstration of it before 1630, and probably not until 1636, just before it was sent to the printer. His own faith in it cannot, therefore, have been purely mathematical, though this has become the fashionable belief among historians by reason of its form of presentation in *Two New Sciences*. Most of the theorems he published there had been derived before Galileo left Padua in 1610; yet even in his 1632 *Dialogue* (completed in manuscript in 1630), he could offer only what he called a probable argument of the times-squared law. The mathematical device (one-to-one correspondence between members of infinite aggregates) required for proof of his Theorem I had been forced on Galileo by the logic of his measurements, not the other way around.[21]

The fashionable view of Galileo as a scientist whose physics was purely mathematical was not his own view, which was this: The *certainty* of conclusions is mathematical, but their applicability to experience is limited by actual measurements. Science remains above all a discourse about the sensible world, with mathematics as its language. The fashionable contrary impression has arisen from the way in which Galileo presented his science with a view to convincing others. Actual measurement there entered not as its beginning but as its confirmation. Hence Galileo's original procedures in experimental measurement were not described, for a good reason still prevalent in science. We describe procedures that anyone can follow with success within stated limits. Galileo described an experiment in which distances along an inclined plane were first measured in given ratios, which can be done by anyone, and then the times of their traversal were measured by the starting and stopping of flow of water from a large container through a small pipe.[22] Thomas Settle, who repeated the procedures in 1960, found that with a little practice he could attain an accuracy within much better than the one-tenth second that Galileo asserted.[23] Musical beats had given Galileo even better results, but he could not guarantee those to everyone, nor could he be sure that others would measure distances accurately in small units. In fact, description of his original procedure would invite scepticism rather than conviction. In several other instances he simplified experimental procedures from the original forms in which he had employed them.

Use of measured water flow to get accurate ratios of times shows Galileo's understanding of actual fine measurement. One of his friends worked at the determination of a seconds-pendulum, a device which Galileo supplied to a former pupil for use in large-scale hydraulic studies. But Galileo replied to his friend that it was not good to use in accurate studies, because there was always an error and it was cumulative. For ratios of short times, water flow was better, and for individual measures in seconds, it was best to use a heavy pendulum of arbitrary length whose oscillations had been counted through a full astronomical day.[24]

IV

In 1604, just as Galileo was writing his first attempted derivation of the times-squared law, a new star appeared. Galileo observed it, corresponded with astronomers in other cities, and gave public lectures in which he explained measurement by sighting and triangulation, and from the lack of detectable parallax in the new star showed that it must be far beyond the moon. A leading professor of philosophy disputed this, arguing that terrestrial measurement procedures did not apply to celestial distances, that astronomers were unable to be certain of the moon's parallax, and that because the new star exhibited change it could not be made of the quintessence of celestial materials but must be of elemental material below the moon. To this Galileo replied in a burlesque dialogue between two peasants: 'What has philosophy got to do with measuring anything? It's the mathematicians you have to trust, and they measure the skies like we measure a field. They don't care if a thing seen is made of quintessence or polenta; that doesn't change its distance.'[25]

Philosophy *had* never concerned itself with actual measurement. Mathematicians had done so, and from 1604 on Galileo, having penetrated into the science of motion by experimental measurements, forsook natural philosophy in favour of sciences based on measurements rather than substance and essence; on laws rather than causes. In 1609 the telescope put into his hands a means of new astronomical measurements, including the orbits and periods of Jupiter's satellites, the period of the sun's rotation, and an enormous improvement in the measurement of visual diameters of planets and apparent diameters of stars. Such measurements were crucial in changing his previous preference for the Copernican system into the conviction that the earth moves. An example must suffice here.

The visual diameter of Jupiter was crucial to a whole set of measurements by which Galileo determined the motions of Jupiter's satellites, a task which Kepler himself had deemed impossible. Starting from an estimate of $1'$ of arc, Galileo improved this to $50''$, and then in 1612, making use of the magnifying powers of his telescopes, he measured Jupiter's disc at two difference distances from the earth as $41\frac{1}{2}''$ and $39\frac{1}{4}''$, very close to modern determinations. At this time he also devised a kind of micrometer for measuring the elongations and separations of Jupiter's satellites. Cross-hair micrometers were impossible with his lens system, but Galileo superimposed the satellites on a scale graduated in units of radii of Jupiter by observing them through the telescope while focusing his other eye on the illuminated scale.

Planetary discs were measurable, unlike fixed stars. Galileo attempted to make the latter experimentally measurable by stretching a thread between his eye and Vega, and moving his eye until the star was just covered. The angle subtended was then given by dividing the thickness of the thread by its distance to the eye, with correction of that distance for refraction in the eye. That in turn was determined experimentally by using a white strip of paper affixed to a wall, and a black strip of half the width held midway between it and the point of convergence of the lines of sight touching the edges of the two strips. The distance from that point to the point at which the eye saw the black strip as just covering the white gave the correction. To measure the thickness of the thread, Galileo took many adjacent windings and divided by the number of windings. In this way he determined the visual angle of Vega as $5''$ of arc, which was 1/24th of the standard estimate by astronomers, who had been deluded by the spurious effect of bright light in a dark field.[26]

Of prime importance to scientific measurement is the theory of errors, initiated by Laplace much later. A step in that direction was taken by Galileo at the beginning of the Third Day of the *Dialogue*, and another in an unpublished paper on errors of estimation. In the *Dialogue* he exhibited a method of utilizing disparate measurements of the same phenomenon made in various places, with instruments of varying sensitivity by observers of differing skill, in a matter of very precise measurement. Galileo's determination by comparing the aggregate adjustments necessary to bring all data into consistency with a definite position for the observed star revealed absurd flaws in the arguments of the astronomer who had selected the observations as showing a very different position.[27]

In connection with measuring the weight of air, Galileo suggested repeated experiments to assure the result. The only result he cited (from memory) was wrong by a factor of two, the procedure being complicated and delicate. In his last book he offered a simplified procedure in which water was forced into a leather bottle by a syringe and weighed with the compressed air, after which it was weighed again with that air released. When it was objected that the operation would have to be carried out in a vacuum to be precise, Galileo replied that one need only add a computed weight of air equal in volume to the counterweight employed to balance the two weighings.[28]

The weight of air was stated relative to the weight of water of the same volume. A different kind of problem was involved in measuring the force of percussion. Galileo approached this in different ways, one of which was to compare the deviation of a bowstring by a dead weight with its deviation by the same weight dropped through a given distance, and then to repeat this with bows of different strengths. All he found out was that the measures implied an infinite force of percussion; that is, that dead weight and impact were incommensurable. The process of seeking actual measures, whether or not successful, was part of Galileo's science. Measures of the resistance of different mediums to speed of descent are another example. Galileo attempted to measure the speed of light at distances up to a mile, by practising with a friend the uncovering of a lantern immediately on seeing light from a distant lantern. Failing to detect a lag, he suggested the use of telescopes placed on widely separated peaks, rather than asserting that light moves instantaneously.

Positive assertions about natural phenomena not backed by accurate measurement are rare in Galileo's science, though of course the backing was often indirect, through mathematical demonstration. This could not reduce certainty, but neither could it replace actual measurement. The nature of Galileo's science is frequently misrepresented, or represented as filled with childish inconsistencies, by writers who either are unaware of Galileo's incessant quest for measurements or mistake actual measurement with what they call 'mathematical precision'.[29] It seems to me that they speak as if Galileo did not believe the real position of a ball on an inclined plane to be where he saw it at a given instant, but somewhere else that could be determined by applying the law of fall, or the positions of Jupiter's satellites to be where his tables placed them and not where he saw them.

Galileo went to some trouble to explain, in various places, that actual measurements are subject to various conditions which are neglected mathematically and are taken into account only when experience shows that to be necessary.[30] Connecting mathematics with a material phenomenon, he said, requires us to know how to balance our books.[31] That was a pretty fair forecast of nineteenth-century physics. I think Galileo made it not because of Plato, or Aristotle, or Archimedes, or Oresme; but because of his interest in and his talent for actual measurements and his confidence that Nature's books must balance for a truly qualified accountant.

Notes

1. Reverence for the cosmological ideal is supposed to have induced mathematical astronomers to retain uniform circular motion, even if the earth could not remain at the centre. Yet the central position for the earth was no less part of the cosmological ideal. More probably that was sacrificed by mathematicians because with nothing more complicated than a table of sines or chords, uniform circular motion permitted easy calculation of positions as seen from any point at past and future times. Non-uniform or non-circular motions would have enormously complicated calculations.

2. 'Again, the physicist will in many cases reach the cause by looking to creative force; but the astronomer, when he proves facts from external conditions, is not qualified to judge of cause. . . . Sometimes he does not even desire to ascertain the cause, as when he discourses about an eclipse; at other times he invents by way of hypothesis, and states certain expedients by the assumption of which the appearances will be saved.' Geminus, cited by Simplicius, ca. 550 A.D.; cf. Thomas Heath, *Aristarchus of Samos* (London, 1913), p.276.

3. Astronomy was in fact classed among the *mathemata*, but in appraising the place of actual measurement in antiquity, Aristotle's general dichotomy between practical and scientific knowledge would hold, only the purely mathematical part of astronomy being strictly scientific, and even that part disassociated from causal investigation by Geminus and Ptolemy.

4. Tycho Brahe so greatly improved naked-eye instruments and measurements that Kepler's discovery of elliptical orbits did not have to await the telescope, which immediately afterwards opened a whole new era of astronomical measurement.

5. This statement is sometimes cited (incorrectly) as if it had been Galileo's own; I have not traced its primary source, which was probably around the turn of the present century.

6. As now practised, scientific measurement is a complicated network of actual measurements in defined units by specified procedures, and hypothetical measurements related to them by established laws and theories. Like physical science as a whole, scientific measurement exists as a stage in a historical process of successive approximation, not as a completed and unalterable whole. Dimensional analysis, for example, is an important part of it, and the theory of errors is another, but neither of these was even formulated until scientific measurement was already far advanced, nor could either have been logically anticipated earlier by even the most profound philosopher. On the contrary, profound philosophers have jeered at conclusions drawn from scientific measurement from Plato to the present as products of naive realism or superficial positivism.

7. See W.A. Wallace, 'The Enigma of Domingo De Soto', *Isis*, 59:4 (1968), 384–401, and his subsequent papers. Father Wallace's fundamental researches into the purpose and method of medieval classifications of motion are indispensable to a clear understanding of the nature of theoretical measurement.

8. A notable example is Thomas Bradwardine's rule relating speeds to paired forces and resistances without definitions of those terms. The rule led to a remarkable mathematical theory of proportion at the hands of Nicole Oresme, but contributed nothing to the study of natural phenomena, which cannot be certified by arbitrary relation of intuitive quantities left not only unmeasured but undefined.

9. It will be seen presently that Galileo's first actual measurement of uniform motions had to await his correct understanding of speeds in accelerated motion, contrary to what might be expected. How such measurements entered into his astronomical conclusions and his view of science generally lies outside the scope of this paper, beyond a few illustrative examples.

10. Conjectures also enter into the reconstruction, but of a different kind from those which have sought philosophical sources for the laws of motion. Chronologically arranged, Galileo's working papers show that a few experimental measurements resulted directly in many theorems. Such an arrangement, in reduced facsimile, is in the press as two monographs supplemental to the *Annali dell'Istituto e Museo di Storia della Scienza di Firenze*.

11. Descartes replaced Aristotle's spheres with vortices of subtle matter, and Cartesians opposed Newton's universal gravitation as an occult power though it was supported by countless actual astronomical measurements.

12. This probably accounts for Galileo's having withheld his *De motu* from publication, its rules for speeds along inclined planes being as badly out of accord with actual tests as were Aristotle's rules for speeds in free fall which Galileo had rejected in the same treatise. See I.E. Drabkin and S. Drake, *Galileo On Motion and On Mechanics* (Madison, 1960), p.69. Galileo's use of theoretical measurement is exemplified on p.25, and of thought-experiments on pp.27–8.

13. The entire letter is translated in S. Drake, *Galileo at Work* (Chicago, 1978), pp.69–71 (hereafter cited by title only).

14. Little discrepancy between mathematics and experience existed in the science of statics based on the lever law. Galileo had tried in *De motu* to make the balance serve as the basis for a science of motion, but neglect of acceleration limited his success. It was that neglect rather than 'material impediments' which truly vitiated his conclusions for inclined planes in *De motu*. This was rooted in Aristotelian reasoning that weight was the cause of speed.

15. Galileo had tried to measure impact by dead weight; see *Galileo At Work*, p.72. Though he found them incommensurable, he had learned experimentally that impacts of the same weight were proportional to distances fallen, which later misled him as explained below.

16. The word 'speed' was difficult to define, though easy to apply without risk of misunderstanding except when acceleration was involved. In the form 'velocity' it had the character of most words ending in -ity, or -ness, created at will to advert to things whose existence in nature has no guarantee beyond our ability to communicate.

17. Details of the ensuing experimental measurement are given in S. Drake, 'The Role of Music in Galileo's Experiments', *Scientific American* (June 1975), 98–104. Other papers cited below were written before this measurement was reconstructed and dated, which should be kept in mind when they are consulted. Historical analysis, like scientific measurement, exists only in stages of successive approximation.

18. Analysis of the document in S. Drake, 'Galileo's Discovery of the Law of Free Fall', *Sci. Am.* (May 1973), 84–92, was amended in *Galileo At Work*, pp.91–3.

19. The letter and the associated attempted demonstration were translated in S. Drake, 'Galileo's 1604 Fragment on Falling Bodies', *Brit. Jrnl. Hist. Sci.* 4 (1969), 340–43.

20. See S. Drake, 'Galileo's Experimental Confirmation of Horizontal Inertia', *Isis*, 64 (1973), 291–305.

21. See S. Drake, 'The Uniform Motion Equivalent to Uniformly Accelerated Motion from Rest', *Isis*, 63 (1972), 28–38.

22. Galileo, *Two New Sciences*, tr. S. Drake (Madison, 1974), pp.169–70 (hereafter cited by title only). Alexandre Koyré held this experiment to be incapable of yielding the results asserted by Galileo, partly because Koyré mistakenly supposed these to include determination of the gravitational constant in free fall; see A. Koyré, 'An Experiment in Measurement', *Metaphysics and Measurement* (London, 1968), pp.89–117. Originally published in 1953, the paper supported the fashionable view that Galileo's science was based on theoretical measurement and thought-experiments alone.

23. T.B. Settle, 'An Experiment in History of Science', *Science*, 133 (1961), 19–23. This paper was not mentioned in the 1968 reprinting of Koyré's 1953 article.

24. See *Galileo At Work*, pp.399–400.

25. Abridged from S. Drake, *Galileo Against the Philosophers* (Los Angeles, 1976), p.38.

26. Cf. Galileo, *Dialogue Concerning the Two Chief World Systems* (Berkeley, 1953), pp.360–64 (hereafter cited as *Dialogue*).

27. *Dialogue*, pp.280–318.

28. *Two New Sciences*, pp.82–6.

29. 'In all physical observations and measurements, one meets with threshold values which cannot be passed, which determine the extreme limits of possible exactness of lengths which have been measured and expressed in millimeters. Statements beyond this limit have no meaning, and are an evidence of ignorance or of attempted deception. . . .

Since mathematics of approximation alone plays a role in applications, one might say, somewhat crassly, that one needs only this branch of mathematics, whereas the mathematics of precision exists only for the intellectual pleasure of those who busy themselves with it, and to give valuable and indeed indispensable support to the development of approximation.' Felix Klein, *Elementary Mathematics from an Advanced Standpoint*, tr. E.R. Hedrick and C. Noble (London, 1932), pp.35–6. These statements by an eminent modern mathematician epitomize measurement in Galileo's science; concerning attempted deception, see *Dialogue*, p.296, and with regard to practical considerations, cf. *Two New Sciences*, pp.223–4.

30. The most complete explanation is in *Two New Sciences*, pp.222–5.

31. *Dialogue*, pp.207–8. The celebrated 'book of nature' passage in Galileo's *Assayer* of 1623, usually quoted out of context, should be compared in context with this later statement.

John Ridley and the South Australian 'Stripper'

L.J. JONES

Introductory Remarks

In a previous issue of this journal I discussed various early attempts to devise mechanical methods of harvesting cereal crops, and especially wheat crops.[1] A crude attempt at harvest mechanization as early as Roman times was noted (the so-called 'Vallus' of Roman-occupied Gaul), and the reasons both for its introduction and later disappearance were discussed in the light of the available evidence. However, as was also demonstrated in the same article, most modern wheat-harvesting machines have their origins in the work of a number of ingenious British mechanicians of the late-eighteenth and early-nineteenth centuries.[2] The earliest, and perhaps also the most original, of these men lived and worked in the counties on either side of the Anglo-Scottish border, and the culminating achievement of this group was undoubtedly the machine constructed by the Rev. Patrick Bell in 1828.[3] As is now well known, Bell's machine saw useful service for many years on the Bell family properties but was never adopted in any general way by British farmers. Labour in Britain was both plentiful and cheap at that time, and as a result machines were neither needed nor welcome in the harvest field. Eventually other (and in the end perhaps more effective) machines followed in the eastern United States, the most notable being those patented by Obed Hussey in 1833 and C.H. McCormick in 1834. Both designs were later taken to high stages of refinement and were sold in large numbers across the temperate regions of the northern hemisphere.

It should be noted, however, that a central feature of *all* the early British and American attempts at producing mechanical harvesters — including the ultimately successful designs of Hussey and McCormick — was that they were invariably crop-cutting machines. That is, the intention of the designer always was to provide a mechanical replacement for a man wielding a scythe or a reaping hook. Of course the form of the cutters employed often differed considerably between the various machines, yet in every case the basic purpose of the cutters was the same — viz., to carry out an initial 'mowing' operation. Afterwards, other workers then collected, bound and 'stooked' the cut stalks for drying out in the traditional way.

But as was also noted (briefly) in my previous article, a harvest machine of an entirely different type appeared on the opposite side of the world, in South Australia, towards the end of 1843. It subsequently

55

became known as the 'South Australian Stripper', or the 'Ridley Stripper' — the latter name honouring John Ridley who was responsible for its introduction. The most striking difference between this and the earlier British and American machines was the fact that the 'Stripper' included no cutters of any kind. Indeed a feature of its operation was that the wheat stalks remained standing in the field after the machine had passed and collected the grain. It thus represented an entirely new departure not only in mechanical design but also in harvesting technique.

At the same time it is an unfortunate fact that, despite its importance and intrinsic interest, the 'Stripper' has remained little known outside Australia and insufficiently appreciated even within Australia. In addition a good deal of misinformation (and indeed some outright nonsense) concerning it has been published over the years,[4] some of which has persisted to the present day.[5] It therefore seems an opportune time to take a fresh look at an invention — for such it undoubtedly was — which was uniquely suited to the special conditions of Australia and so proved to be of very great importance to the development of the country's wheat industry. The following pages are devoted specifically to that end.

A Labour Crisis in the Colony of South Australia in 1843: the Context for the Invention of the 'Stripper'

The new colony of South Australia was officially proclaimed on 28 December 1836 at what is now the seaside resort of Glenelg by its first Governor, Captain John Hindmarsh. A simple ceremony, held under a large tree, signalled the beginning of an audacious colonization experiment, based on principles which had never before been tested in practice. In fact the colonization venture itself was not one undertaken directly by the British Government. It was, at least in the initial stages, a project planned and executed by a private company formed by a mixed group of wealthy English merchants and social reformers, with little more than the paternal blessing of the British authorities. The common ground of the company directors was a shared belief in certain new colonization theories put forward by Edward Gibbon Wakefield, a leading political economist of the day.[6] Wakefield's theories were designed with two principal ends in view — viz., to create new opportunities for the investment of British capital overseas, and at the same time to provide useful employment for some of the surplus labour of the mother country.[7]

The economic basis of Wakefield's scheme was the sale of large tracts of Crown land (in southern Australia or other suitable virgin areas), in properly surveyed blocks, to British capitalists who wished to take up farming. The types of investors required were those who were prepared themselves to go out to the new region, at their own expense, and become the landed proprietors there. In turn, the proceeds from the sales of land were to be devoted to providing free passages to the new colony for unemployed British labourers and their families. These people would then (Wakefield hoped) take up employment with the new landowners, to the eventual benefit of both. In this way, he believed, a viable colony might be established which would be free of the undesirable features so evident in convict-based settlements such as New South Wales and Van Diemen's Land.[8]

Unfortunately Wakefield's 'Utopia' was not to be realized so easily. During the early years of the colony the public officials continually quarrelled among themselves, the labourers showed little enthusiasm for the rough conditions of work in the raw Australian countryside, and the settler-proprietors entirely misjudged both the nature of the land and its agricultural potential. As a result the local agriculture remained in a feeble state for a very long time, and virtually all the grain and flour needed to feed the population (which by the end of 1839 numbered about 10,000 persons) had to be imported at considerable expense. This general inactivity in fact persisted for several years, and a huge debt for imported foodstuffs was piled up. Finally, early in 1841, the economy of the province collapsed completely. The interacting circumstances leading up to the crash are in themselves a fascinating study,[9] but they are not directly relevant to my central theme and so will not be pursued here.

There were, however, important consequences of this severe predicament for South Australia, in the form of swift official intervention by the British Government. Sweeping changes were introduced with a view to preventing the total collapse of the colony itself. On 10 May 1841 the then Governor (Colonel George Gawler) was summarily dismissed and replaced by a young Army officer, Captain George Grey. In addition, new Acts were prepared and passed through the British Parliament which provided for a much modified style of administration for the colony.[10] The Parliament at Westminster now assumed direct responsibility for South Australia's affairs, and indeed the colony was placed on an administrative footing similar to all the other Australian colonies.[11]

Governor Grey, working under strict orders from the Colonial Office in London, at once imposed stringent new measures designed to force the labourers out of Adelaide and on to the land, and in this he soon succeeded. Agricultural development in the region then advanced rapidly. Large areas of new land were cleared and cultivated, and within only a few months the number of acres sown to wheat almost quadrupled (see Table 1 below). Optimism once more began to pervade South Australian society as the province again became busy and a good season gave promise of a bumper harvest. The figures in Table 1 perhaps indicate best how quickly inactivity was replaced by

TABLE 1[12]

Year	Population	Acres cultivated	Acres under wheat	
1836	546	Nil	Nil	Governor Hindmarsh
1837	3,000	8	Nil	
1838	6,000	86	20	Governor Gawler
1839	10,000	443	120	
1840	14,600	2,503	1,059	
1841	?	6,722	4,154	Governor Grey
1842	?	19,790	13,892	
1843	17,366	28,690	23,000	
1844	19,000	26,918	18,980	
1845	21,759	26,218	18,838	
1846	25,893	33,292	26,134	

real agricultural progress following Grey's arrival and assumption of office.

However the forced shift of labourers to jobs on the land had maximum effect during Grey's second year of office (1842), when the area planted with wheat again more than trebled to nearly 14,000 acres. When that harvest was gathered at the end of 1842, South Australia had become self-sufficient in the production of grain for the first time. Indeed there was a small surplus available for export.

But at the same time the rapid expansion of wheat planting in 1841 and 1842 brought with it a crisis of a new and different kind. By the end of 1842 all the labourers in the province had been absorbed into farm employment, and it now became evident that there was a serious shortage of hands to reap and bring in the crops in the very short South Australian harvest season. (It is important to notice here that the brevity of the harvest period is due to the hot, very dry climate, which causes the grain to over-ripen rapidly and then to 'shed' naturally from the ears.) Fortunately the Governor had anticipated the problem well before the onset of the harvest, and had made arrangements for about 150 soldiers to be sent from New South Wales to help out. Much later, Grey himself remarked that:

> At the pruning hook, in getting in that harvest, they were of vast assistance, and not often have soldiers been more nobly occupied.[13]

In addition to the squad of soldiers, virtually every able-bodied man in the colony was conscripted to take part in gathering the crops. Governor Grey ordered all Government offices to close in order to release as much manpower as possible, and the colony's 'gentlemen', who in the ordinary way would never have participated, shared the work of reaping with the farm labourers. Sailors from ships in the harbour deserted their vessels in order to take advantage of the high wages being offered for harvest work. The Governor himself is reported to have set an example by rolling up his sleeves and taking his turn with the reaping hook. In this way the entire harvest was in the end successfully gathered, but in terms of manpower there was nothing to spare.

A few months later, after the new crops had been planted, it became clear that the labour situation at the next harvest (i.e., at the end of 1843) would be far more serious. The total planting of wheat in 1843 was about 23,000 acres — an increase of almost two-thirds over the previous year — whereas the population had increased only very modestly. Obviously, then, a considerable part of the crop would have to be left in the paddocks and lost unless something could be done quickly to supplement the labour force. Realizing this, Governor Grey petitioned the Home Government for the immediate dispatch of 400 migrant labourers, but to no avail. Only 110 were sent, and in any case none of these arrived in the colony until sometime during the following year (1844).

As events turned out, however, South Australia's harvesting problem was solved, not by imported labour, but by the timely introduction of a machine to replace labourers in the harvest field. This machine of course was the 'Stripper' and, as will be shown later, two such machines were in service before the 1843–44 harvest season ended. Together with the available labourers using traditional hand-reaping tools, the new machines enabled most of the 23,000-acre crop to be gathered successfully, and shortly afterwards the first substantial exports of grain from South Australia were made. Thereafter (as we shall also see later) the 'Stripper' rapidly became the preferred harvesting implement of the region, and as such played a major part in the enormous expansion of the local wheat industry (and of course wheat exports) which occurred over the remainder of the century.

The 'Corn Exchange Committee' and the 1843 Design Competition

By about mid-year of 1843, after sowing had been completed, the full gravity of the situation to be faced at the coming harvest at last began to be appreciated by the settlers generally. Indeed the more thoughtful colonists and public officials were able to see as well the wider, much more serious implications, namely that there was a real prospect of the future agricultural growth of the colony being drastically curtailed, or perhaps stopped altogether. Opinion quickly favoured a search for some form of harvesting machine, perhaps operating along similar lines to Bell's or one of the other existing overseas machines,[14] to supplement the now quite inadequate local labour force. Soon there was a strong body of opinion that mechanical reaping represented the *only* feasible solution, not only to cope with the immediate crisis, but for the long term as well.

At the same time, however, there was some disagreement concerning the approach most likely to satisfy South Australia's particular needs. As pointed out by a correspondent to the Adelaide press on 26 August of that year, some farmers still hoped that men using scythes would prove sufficient in the end, whereas:

> others are thinking of the construction of a machine something after the fashion of Bell's reaping machine, of which there is a plate and a description in Loudon's *Encyclopedia of Agriculture*.[15]

The writer then went on to say that whereas he himself lacked the requisite knowledge actually to build a suitable machine:

> there is a good deal of unemployed mechanical skill and ingenuity in the colony, [and] it might not be amiss to call attention to this subject. Any man who should construct a machine which would answer this purpose would be a benefactor to the colony.[16]

Unknown to the writer, however, at least three of his fellow colonists had already begun the construction of machines of their own devising, and according to the Adelaide *Observer* a week later the number could have been as high as five or six. Unfortunately no further details were given.

Coincidentally, on that same day (2 September 1843) a quite prophetic letter appeared in the columns of the rival paper, the *South Australian Register*. On this occasion the author identified himself only as 'A Wheat Grower', but a few weeks later it was revealed that he was really John Ridley, a flour-miller from Northumberland who had arrived in Australia three years previously and had set up the colony's first flour-mill just outside Adelaide. His letter read in part as follows:

> Gentlemen,
>
> After what has been accomplished by mechanical science we need not fear over-rating its powers in asking assistance in reaping our grain crops, without which they are in danger of being of questionable value — reaping and threshing threatening to absorb the whole proceeds. As it is not upon the grain itself the labour is expended but upon the straw, could we have a machine to collect the heads and leave the straw standing, threshing and winnowing the grain at the same time, it is evident that a large proportion of the labour at present employed would be saved. . . . a projecting row of prongs with cutters may be made to collect and detach the heads of grain. By taking a breadth of five feet two acres may be gone over in one hour at the rate of travelling of three miles per

hour, and instead of two pounds per acre two shillings will place the grain in bags.

I throw out these hints under the hope of attracting the attention of the readers ... I do not think the mechanical ingenuity of the Province could be better employed than in bringing something of this sort to the test of actual experiment.

Yours Truly
'A Wheat Grower'[17]

The references in this letter to 'cutters' and to the possibility of performing all three harvest operations at once in the same machine are of some interest, since the machine actually introduced by Ridley eleven-and-a-half weeks later in fact met neither criterion. It included no cutters of any kind, and was also without a winnowing function although, as we shall see, it *did* combine the other two basic operations of reaping and threshing. Clearly Ridley had not fully crystallized his ideas at this earlier stage, but equally there can be no doubt that he foresaw the need for a radical change in harvesting technique in the region if South Australia's difficulty was to be satisfactorily overcome.

In the meantime, however, others in the colony had taken steps of a different kind to meet the coming harvest problem. In late August or very early September 1843 the 'Corn Exchange Committee' was formed in Adelaide by a group of prominent local men who were in the habit of meeting regularly for social purposes at Payne's Hotel in the town. In general these individuals were not especially expert in mechanical matters. Indeed most were either local merchants, businessmen or farmers. Their aim in setting up the 'Corn Exchange Committee' was nevertheless a sound one — to provide a forum for public discussion, and to try to encourage those with inventive talent and/or ideas for mechanical reaper designs to come forward. In this way, the members hoped, something might emerge which could be of help at the coming harvest. John Ridley was not a member of the Committee despite his by now acknowledged mechanical expertise. Probably this was because, being a strict Wesleyan and teetotaller, he did not frequent Payne's or for that matter any other hotel.

Whatever else may be said about it, the 'Corn Exchange Committee' was an active body. During September it sponsored public meetings in Adelaide at which all those who had plans for reaping machines, or were otherwise interested, were invited to debate their own and other designs and to discuss the possibility of manufacture. In addition the Committee appointed a judging panel to evaluate the designs submitted, and also offered a small prize for the winning entry. Four meetings were held in all — on 12, 19, 20, and 21 September respectively — at which about eighteen or nineteen persons[18] came forward with drawings, models, or just ideas for discussion. All four sessions were also very well attended by the general public, whose interest and concern by now had been thoroughly aroused.

John Ridley was one who met the examining panel (at the final meeting on 21 September), although he made it clear that he was *not* an entrant in the Committee's contest. He had come, he said, merely to acquaint the Committee of the fact that he was already engaged in constructing a full-sized machine of his own design in the workshops attached to his flour-mill at Hindmarsh, and that the project was well advanced. His machine was designed, he also told the meeting, to carry out 'the subsequent process of thrashing as well as reaping'.[19] The Committee and the general public would be welcome, Ridley said in conclusion, to witness its trial in the field in the near future.

Of the proposals actually submitted to the judging panel over the four meetings, all were 'cutting' machines of one kind or another, mostly slight variants of earlier overseas designs such as Bell's, Smith of Deanston's or McCormick's — with one notable exception. This was a novel proposal presented (in the form of a working model) by Mr J.W. Bull of Mount Barker, which was described in the Adelaide *Observer* a few days later as follows:

> His [Bull's] machine consisted of a long-toothed comb, fixed to the back of a close-bodied cart; the teeth being operated upon by four revolving horizontal beaters, with square edges, which would have the effect of taking off the ears of corn and depositing them in the body of the cart, a wide bag being placed under the after part of the comb to catch any dropped grain.[20]

In his supporting remarks to the Committee, Bull stated that 'I do not think that we want or can afford to pay for a nice mode of in-gathering corn', and then went on to discuss ways of utilising the straw and spilled grain left in the field. These, he said, could serve to fatten sheep and pigs, thus recouping the waste part of the wheat crop in another form.[21]

From this it is obvious that Bull had not grasped the true potential of his design at this stage. He expected, mistakenly, that because of its simplicity his machine would do no more than roughly harvest the grain. Whether because the members of the judging panel concluded likewise, or for some other reason, Bull's scheme did not appeal and it was passed over without further ado. In so doing the panel not only misjudged the situation entirely, but they also missed a very promising opportunity since, of all the designs submitted, Bull's was the only one which properly took into account the special local conditions. Had the panel members been more knowledgeable they might have realized, as Bull himself had done, that cutting was fundamentally a wrong approach to harvesting crops which were as dry and as brittle as those which matured quickly in the severe summer conditions of South Australia. However, ignore Bull's scheme they did, and after apparently careful deliberation the examiners chose a design from a Mr B. Swingler as the best of the entries.

Over following weeks an attempt was made to construct a prototype Swingler machine, but this had to be abandoned when it became evident that there were a number of very serious faults in the basic design. A programme of construction for the next most favoured entry — one by a Mr A.J. Murray — was then begun in its place, but this also came to nothing. Eventually, on 3 November 1843, an Adelaide practical 'mechanic' named Samuel Marshall became the first actually to test a reaping machine of his own design and manufacture in the field, and once more it was a machine of the 'cutting' variety. His cutting apparatus worked admirably, but unfortunately another mechanism intended to throw the cut stalks into bundles for later binding proved a total failure. After several attempts to rectify the trouble a disappointed Marshall was finally forced to acknowledge that several features of the gathering mechanism were basically unsound, and so he returned quietly to his workshop to redesign and replace it.[22]

This then was the situation when, twelve days later, an announcement in the Adelaide press took the entire community by surprise. This was that John Ridley had now completed his machine, and that a public trial held at Wayville the previous day (14 November) had been triumphantly successful. In front of a large crowd of onlookers, the 'Stripper' (as it came to be called later) had harvested the grain faultlessly and very efficiently for the whole of the afternoon. Those who were there, said the press, had come away convinced that this machine would solve the colony's harvesting problem once and for all.

During the following weeks, when the harvest proper got into full swing, this opinion was fully confirmed. As will shortly be shown, this machine and another hurriedly constructed at Hindmarsh by Ridley's men made an invaluable contribution at that harvest. As a result, farmers in the region quickly lost interest in the 'cutting' machine proposals which had been considered previously by the 'Corn Exchange Committee'. Indeed it is probably fair to say that every one of these had been virtually forgotten by the end of the year.

Perhaps, however, it should also be mentioned here that the operating principle of Ridley's successful machine seems to have been markedly similar to that proposed a few weeks earlier by J.W. Bull of Mount Barker. As noted previously, Bull presented a design in model form at the 'Corn Exchange Committee''s second meeting on 19 September, but failed to arouse any particular interest among the examiners on that occasion. Curiously, neither the press nor anyone else appears to have remarked upon this similarity when Ridley's machine was first introduced, and even Bull himself failed to draw attention to it publicly until some sixteen months later.[23] Subsequently, however, a protracted and rather unpleasant controversy developed over which of the two men should be credited with the machine's invention, and indeed it is still an interesting question. It will be further discussed in a following section.

Figure 1. John Ridley's original 'Stripper' — a sketch by 'N.R.F.' in 1845. [Reproduced by Marcus Collisson, *South Australia in 1844–45* (1845).]

The Introduction and Astonishing Success of Mr John Ridley's 'Locomotive Thrashing Machine'[24]

In the two weeks or so preceding the first public demonstration of Ridley's new harvest machine at Wayville, the local newspapers carried somewhat contradictory reports of some *private* trials evidently carried out by Ridley immediately after construction of the machine was completed. According to one paper these had shown the machine in its first form to be a complete failure.[25] According to the others, however, only 'slight modifications' were needed to make it workable.[26] Testimony given many years later by John Dunn (Ridley's foreman in the workshops at the 'Hindmarsh Flour Mill', where the machine was made) showed that in a sense both stories were correct.[27] Apparently trouble was experienced with the original (wooden) gathering comb. According to some reports the wooden teeth warped with moisture, but other evidence suggests that the difficulty was caused principally by the teeth having an incorrect cross-sectional shape. Either way, the Adelaide *Observer* made it clear that at first the machine tore up part of the crop by the roots, and tended to pass over the rest without collecting the grain at all.[28]

As will be shown in more detail later, a solution was soon found by John Dunn and a fellow workman at Hindmarsh, John Dawkins, who made and fitted a new comb made of iron and having a better tooth shape. It was in this form that the machine appeared at Wayville on 14 November for its initial public showing, on which occasion, as already noted, it was outstandingly successful. Indeed it ushered in a revolution in wheat harvesting in South Australia.

Immediately after the Wayville demonstration the three Adelaide newspapers were once more in complete agreement. The *South Australian Register* was the first into print on the subject (the following morning), stating that:

> Mr Ridley's reaping and threshing machine ... on being tried yesterday, was found to answer admirably. The success, in fact, is most triumphant, and Mr. Ridley is entitled, not only to the thanks of the colonists, but to those of the whole agricultural world. The two separate processes of reaping and threshing at the same moment are complete.[29]

Equally complimentary reports soon followed in the other newspapers. The *Southern Australian* commented particularly on the extraordinary thoroughness with which the machine gathered up the grain, despite the fact that the crop on which it was tested was not yet fully ripe. Indeed this reporter went so far as to predict that:

> under proper management there need not be the loss of a bushel of wheat in twenty acres.[30]

All three newspapers emphasised the simplicity of operation of the machine, and also its highly unusual feature of collecting only the grain whilst leaving the stalks standing in the field after it had passed. That is, the crop was 'stripped' instead of 'reaped' and, as pointed out in the *South Australian Register*'s report, the traditionally separate operations of reaping and threshing were accomplished simultaneously. Because of this, the *Southern Australian*'s writer said,[31] the machine strictly speaking was not a reaping machine at all, as that term was commonly understood. The same point was made by the Adelaide *Observer*'s reporter when he referred to it as:

> a locomotive threshing machine ... [employing] ... a series of protruding knives with there [sic!] edges downward, and having the appearance of a bent steel comb upon a large scale ... [by which] ... the ears of corn are collected and brought under the operation of the drum.[32]

Both papers asserted that 'with a fair crop' the machine could be expected to thresh out the wheat at a rate of more than an acre every hour. According to the *Southern Australian*[33] this could be worth the labour of 'fifty or sixty good men' to the farmer. However the Adelaide *Observer* reiterated its earlier claim that the machine had in fact failed when first tested, and now acknowledged only that 'a further trial has [since] established its complete success'.

According to the *Southern Australian* the first successful demonstration of the new machine took place 'on a farm near Bowden, belonging to Mr. Ridley, at present possessed by a person named

Langman'.[34] However a later student of South Australian history, the
Rev. William Gray, gave the location as:

> a field where Wayville is now, Section 221. The land belonged to
> the South Australian Company, rented at the time by Barrow and
> Gouger.[35]

The majority of other accounts of the event indicate that the
location given by Gray is the correct one. Most likely the *Southern
Australian*'s reporter confused the public trial of the machine with the
earlier private tests carried out by Ridley not far from his Hindmarsh
workshop, after which the original wooden comb was replaced.

When, on the afternoon of 14 November 1843, the news spread that
Ridley's machine was working successfully, a crowd of spectators soon
converged on the scene. By mid-afternoon several scores of people had
arrived, and they all watched in astonishment as 'the heads of corn
were thrashed off perfectly clean'.[36] The driver of the machine on this
occasion was one Thomas Munroe, and he in turn was directed from
the ground by John Wilkinson. Both were in John Ridley's employ at
Hindmarsh at the time. Strangely, none of the reports mentioned
Ridley himself at the scene, or for that matter John Dunn and John
Dawkins, both of whom were also Ridley's employees and had carried
out the major part of the actual construction of the machine. However
it seems unlikely that any of the three would have stayed away from
such an important event.

The Manner in which Ridley Conceived the 'Stripper' Design

As already described, during September and October 1843 public
interest in the colony had been focused on the general subject of
reaping machines, but not especially on any one design. Ridley's plans
had received some press attention along with the rest, but no firm
details of what was going on in his Hindmarsh workshops had been
made public. The activities of the 'Corn Exchange Committee' had
engaged the attention of the settlers almost exclusively, and Ridley's
project had progressed virtually unnoticed. Ridley himself also
contributed to this state of affairs, since he preferred to keep both his
ideas and news of any progress made to himself until he actually
accomplished what he had set out to do. Apart from the workmen he
employed at Hindmarsh and a few close friends, no-one seems to have
known what kind of machine Ridley had in mind until it appeared at
Wayville on 14 November.

Ridley also said nothing *after* the Wayville trial about how he came
by the basic ideas for his design in the first place. Neither did he
indicate how these had been developed and refined before he began the

task of building a machine for testing. In fact he made no public statement on any such matters until 1886, almost at the end of his life, and even then he gave little away. Had he been more forthcoming in this respect the hurtful controversy over rights to the invention (briefly mentioned earlier) might never have arisen.

However it is also true that Bull made no public mention of the similarity in operating principle between Ridley's working machine and the model he himself had displayed to the 'Corn Exchange Committee' in September. Indeed it was not until March 1845 — that is, sixteen months after what most would regard as the appropriate time — that Bull first laid claim to the discovery of the 'stripping' principle,[37] and thus by implication to the invention of the 'Stripper' itself. This is most curious, since as time went on Bull hinted more and more broadly, and eventually stated explicitly,[38] that Ridley was guilty of stealing the idea from him. During the 1870s and 1880s Bull gathered considerable support for his case, but he was never able to back up this accusation with anything like acceptable proof. Nevertheless, by diligent lobbying he did finally achieve a measure of official recognition when, in September 1882, he was awarded a cash bonus of £250 by the South Australian Parliament 'for services in improving agricultural machinery'.[39] Whether an award in these particular terms was appropriate is of course arguable, since Bull's contribution, if any, was in no way connected with making improvements to existing machines.

John Ridley
(1806–1887)

John Wrathall Bull
(1804?–1886)

Figure 2. The principals in the dispute over rights to the invention of the 'Stripper'.

But in any case the award failed to end the dispute either then or later. In fact it has been revived at intervals ever since by supporters on both sides, and it is still a subject capable of generating a sharp debate in some quarters.

Ridley himself made only one public statement on the matter, in a letter written in his old age and published in the Adelaide press in May 1886.[40] In this he flatly denied having been influenced by Bull's proposal, or indeed any other, at the time he developed his own machine. The initial inspiration had come to him, he said, after he had noticed an account of 'an old Roman invention' in Loudon's *Encyclopedia of Agriculture*. Otherwise he had received 'not the least help or suggestion' from any other source whatever.

The Roman invention referred to was of course the 'Vallus'[41] but it should be remembered that this was not a 'stripping' machine in the modern sense. The mode of operation of the 'Vallus' was to snap off the whole heads of grain after these had become wedged in the projecting frontal spikes. It did *not* remove the grain from the ears as did Ridley's machine, and so the collected grain-heads had to be threshed afterwards to separate the grains from the husks. Further, the 'Vallus' had no revolving beaters or other apparatus for striking the trapped heads of grain (see Figs. 1 and 3), but relied solely on its own forward motion to detach these from the stalks. As far as is known Ridley never explained how he conceived the idea of adding mechanically driven beaters above the comb to accomplish the threshing operation as well.

Apart from Ridley's own testimony just noted, John Dunn provided by far the most authoritative account of how the original 'Stripper' design evolved. Dunn was in Ridley's employ at the time, as he afterwards recorded in his published reminiscences:

> I was with Ridley as Engineer from October 1842 till some time in 1844, and during that time ... [I] helped to build his first reaper, and was the first to get the machine to work rightly in the field.[42]

In fact Dunn seems to have been Ridley's foreman in the workshops at Hindmarsh when the first 'Stripper' was built there, and so what he had to say on the subject generally should receive close attention. Long afterwards (at the end of 1890) Dunn wrote to the press following the death of John Dawkins, his fellow 'mechanic' at Hindmarsh, and in this he included the following remarks on the 'Stripper''s origins:

> Ridley's first ideas were bills — sharp, shear-like knives to cut close to the ground, and next to cut the heads off and let them fall into a box, ... The next year it was decided to strike the head instead of cutting it off. The comb, now amongst the simplest parts of the machine, was then the most difficult to hit on, the trouble

being in getting the right shape or form. A variety of shapes were tried. At last we got one that did fairly well.[43]

Elsewhere Dunn referred to the source of Ridley's inspiration and its later practical realization as follows:

> The idea of the Ridley reaper was obtained by the person from whom it took its name, from a sketch in an old book lent him by Dr. W.J. Browne. The sketch represented two Egyptians gathering in the harvest; one of them wheeling a barrow and the other knocking the heads containing the grain into it with a stick or some simple hand instrument. This is the principle of the Ridley reaper, ... The first machine was built by Mr. John Dunn and Mr. Dawkins, who were in Mr. Ridley's employment .... It was sometime in the early part of 1843, I think, that Mr. Ridley conceived the idea of striking the heads instead of cutting them off, and explained his original notions to me and others in the workshops. I worked a good part of the year altering the machine in various ways, but it proved a most obstinate baby.[44]

By the time Dunn wrote down these comments he was over eighty years of age, and more than forty years had elapsed since the events concerned had taken place. That being so, some imperfections in his recall of detail might be expected and allowed for. The 'barrow' mentioned above was in all probability the Gallic reaping cart (or 'Vallus'), while the two men attending it might easily have been Gauls instead of Egyptians. In addition, the old book lent by Dr Browne could well have been J.C. Loudon's *Encyclopedia of Agriculture* — the source named by Ridley himself in his 1886 letter to the press.[45] In all other essential respects Dunn's testimony agrees with that given (entirely independently) by Ridley himself. Since both men were at the centre of the events concerned, this agreement takes on an added significance.

It thus seems plausible — indeed likely — that Ridley *did* find a starting point for his ideas on reaper design in J.C. Loudon's well-known *Encyclopedia*, and not in the model displayed by J.W. Bull. That work, after all, provided one of the few accounts of the ancient 'Vallus' then available. John Dunn's evidence shows that while Ridley first contemplated a type of 'cutting' machine (sometime during 1842), he recognized the far greater potential of the 'stripping' approach *early in 1843* — that is, many months before the appearance of J.W. Bull's model.

It is also probable that the feature of the old Roman harvesting method which appealed to Ridley in the first instance was its admirable *simplicity*. This would surely have interested a man of a strongly practical outlook, who was aware also of the conditions to be faced. Ridley knew very well that there was a great scarcity of mechanical repair facilities in the colony, and that mechanically unskilled farm labourers would inevitably find complex machines too difficult to manage.

Neither Dunn nor Ridley gave any indication at all of the source of the latter's ideas for the mechanically driven beaters and enclosing cowl. Most likely, however, the basic notions came from the conventional threshing machines of the day. By the 1840s such machines had been in common use in Britain for several decades. Also significant, perhaps, is the fact that a diagram of the mechanism of Andrew Meikle's threshing machine was included in Loudon's *Encyclopedia*, and in the very same chapter as the description of the Roman 'Vallus'.[46] However the notion of combining a 'Vallus'-like comb with a threshing apparatus, and the concept of directly threshing the ears of a growing crop, were quite new. Both appear to have been pieces of brilliantly original insight on Ridley's part. Certainly it was these features which clearly distinguished his machine from the ancient 'Vallus', and which distinguished Ridley's (and Bull's) approach to harvesting grain from all others employed up to that time.

The Mode of Operation of the 'Stripper', and its Particular Suitability to South Australian Conditions

The earliest press reports of Ridley's machine made much of its unusual operating principle, its simplicity and its remarkable efficiency in collecting the grain. Also, all three local newspapers were unanimous in their opinion that it was the answer to the colony's harvesting problem. At the same time, however, there were some significant discrepancies between their separate descriptions of the machine's capabilities. For example, the Adelaide *Observer* claimed that it would 'collect the ears, thresh them out, winnow the corn and fill the bags'[47] whereas neither the *Southern Australian* nor the *South Australian Register* made any mention of winnowing or of bag-filling. Later and more detailed accounts confirm that the *Observer*'s statements were entirely erroneous, and that a final winnowing process was always necessary after the wheat was emptied from the machine. Bagging was also done afterwards, and again *outside* the machine.

A further difference of opinion over the 'Stripper''s mode of operation becomes evident when the account of F.S. Dutton, in his book of 1846,[48] is compared with that of Anthony Forster, published twenty years later.[49] Forster claimed that the heads of grain were *broken off* preparatory to being threshed by the revolving beaters. Dutton, on the other hand, indicated that the grain was threshed directly from the ears whilst these were held momentarily in the teeth of the comb and still attached to the stalks. A careful reading of the early newspaper reports of the machine shows that Dutton's account is the correct one

— that is, the beaters operated directly upon the growing crop rather than requiring the heads of grain to be detached first. No doubt a few heads did actually break away when they became wedged in the comb, but not the majority. Those that broke off probably collected in the concave sheath or cowl below the beaters, where they would then have been quickly broken up. (See also Fig. 3 below.)

The core of Ridley's new harvesting method (and evidently of J.W. Bull's also) was the concept of dislodging the individual grains by applying a sharp blow to the ears. This in turn depended upon the wheat being sufficiently dry to allow the grains to separate from the husks without difficulty. The 'stripping' method was therefore very well suited to South Australian conditions, where rain in summertime was infrequent and high temperatures ensured that the crops were thoroughly dried out by harvest time. The necessary blows upon the ears were supplied by four rapidly moving flat flails (or 'beaters') mounted on a revolving spindle driven from the main ground-wheels. This assembly was positioned above and slightly behind the rear end of the projecting comb, so that the heads were struck at precisely the instant they became wedged in the teeth. The grains thus dislodged were then swept backwards into a cowling below the beaters, and from there upwards and over a 'hump' into a storage compartment at the rear. Chaff, husks and dust were propelled more steeply upwards because of their lighter weight, and flowed out in a stream through a sort of 'chimney' provided at the top. (The chimney is plainly visible, for example, in Fig. 1.)

Apparently J.W. Bull envisaged a very similar arrangement for dislodging the grains of wheat from the whole ears, but intended using a different means for imparting motion to the beaters to that adopted

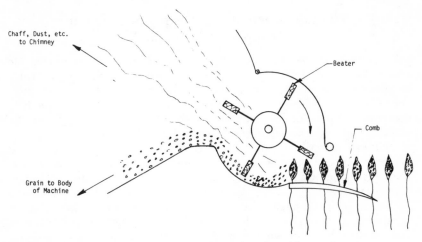

Figure 3. Diagrammatic illustration of the 'comb-and-beaters' arrangement, and of the basic 'stripping' action.

Figure 4. An illustration of Ridley's original 'Stripper' showing the drive to the beaters. [Copy by courtesy of the State Library of South Australia, Adelaide.]

by Ridley. Bull's idea was for a combined chain-and-belt drive,[50] while Ridley preferred a pair of gears and a crossed leather belt (see Figs. 4 and 6 below). This, however, was a difference of mechanical detail only, and not one of basic principle.

The Construction and Testing of the First 'Stripper' at Hindmarsh

John Dunn's testimony as to the time when Ridley first conceived the idea of harvesting by 'stripping' is explicit. He stated that it was 'early in 1843' — that is, many months before the 'Corn Exchange Committee''s meetings were held and J.W. Bull's model publicly displayed. However, no available information states exactly when the construction of the first 'Stripper' was begun at Hindmarsh.

Ridley's letter (signed 'A Wheat Grower', and published on 2 September 1843)[51] mentioned the possibility of combining the traditional operations of reaping and threshing in the same machine. This suggests that Ridley had probably finalized at least his design approach by that stage, but the letter said nothing whatever about construction. A week later the editor of the Adelaide *Observer* claimed that he had been shown the mode of operation of Ridley's design.[52] Once again, however, there was no indication given that the editor had seen a machine actually being built.

All that we can say for certain is that construction commenced well before 21 September, the date on which Ridley reported to the 'Corn Exchange Committee' that his project was 'well advanced'.

There is also some uncertainty over the identities of the men who actually took part in building the first 'Stripper' at Hindmarsh, although some can be positively identified. John Dunn and John Dawkins between them apparently carried out much of the assembly and fitting, and indeed Dunn seems to have acted as 'leading hand' for the entire group of craftsmen involved.

Very much later, on the occasion of Dawkins's death at the end of 1890, Dunn placed on record a tribute to the special contribution which his old friend and workmate had made to the success of that first machine. The 'Stripper' was, said Dunn:

> set to work one year sooner than it would have been but for a casual remark of his [Dawkins's] put into practice within a few days.[53]

Dunn had already described the circumstances in which the 'casual remark' by Dawkins had proved so valuable, in a long letter to the press four years previously.[54] In this he revealed that the original machine had persistently malfunctioned during a series of private tests carried out immediately construction was completed. The machine proved to be, said Dunn, 'a most obstinate baby'. (Clearly it was these difficulties to which the Adelaide *Observer* referred in its earliest notices, when it was claimed that the machine was a failure, tearing up the crop by the roots.) But also according to Dunn, the cause of the trouble was not warping of the wooden comb (as stated in some early press reports), but the shape of the teeth themselves. Apparently Ridley himself also failed to recognize this, since he first tried several other small modifications in an attempt to make the machine work properly, but all without success. According to Dunn, Ridley finally became 'frustrated and ill' and confessed to being 'quite beaten', at which point he went home for the day after instructing Dunn and Dawkins to 'do what you can'.

It was Dawkins who then suggested trying a new tooth-profile for the comb, and proposed a tapered triangular form similar to an ordinary three-cornered file (one of which happened to be lying nearby on the workbench at the time). Dunn agreed, and there and then the two set about making a new comb (this time of iron) to replace the unsuccessful wooden one. What followed was described by Dunn in some detail:

> of course, no iron was available wherewith to make them [the teeth]. Some one had seen some half-inch or five-eighths-inch square [iron] in town. I sent for it, and the smith made a V tube to fit in the anvil. The teeth were forged, two flat iron bars fitted

to bolt up under the machine, 2-inch paling nails put between
each tooth to keep them the proper distance apart, the side bars
pinched together with bolts. Then I turned the teeth downwards
and had lead run in between the bars to prevent the teeth from
shaking out going over the rough ground. This took some days,
and long ones[55] .... We bolted up the new comb and it went off
splendidly. I never saw a machine work better or take off a crop
more cleanly.[56]

From this it is seen that indeed Dunn and Dawkins did play an
important part in bringing Ridley's ideas to practical realization, but
the identities and contributions of the others involved cannot be
determined so definitely. A 'mechanic' named John Wilkinson certainly
had some part in it[57] and, as noted previously, Wilkinson also helped to
operate the machine at its first public demonstration at Wayville. A
certain 'Mr Cooke who was a wheelwright'[58] was another who
contributed, as was also a Mr Langman who worked for Ridley as a
blacksmith at the time.[59] (Note also that in all probability this was the
same Langman mentioned earlier as Ridley's tenant on the farm at
Bowden, where the initial private tests were conducted.)[60] According to
a much later source[61] Mr Langman made the ironwork for the original
'Stripper', and Messrs Burton and Isaac Hall did the woodwork.

There is also direct evidence that Samuel Marshall was involved in
the construction programme, in the form of a letter of his to J.W. Bull
in late 1875. In this Marshall stated that he: 'assisted in the
construction of the first machine that was made at Mr. Ridley's
place'.[62]

However it is known that during the months of September and
October 1843 Marshall was fully occupied with the building of his own
reaper at his own premises in Adelaide. As mentioned earlier, his
machine was first tested in the field (and performed unsatisfactorily) on
3 November — that is, only eleven days before Ridley's notable
demonstration at Wayville. That being so, any assistance Marshall
gave to Ridley at Hindmarsh could only have been in the very last
stages of the latter's construction programme.

From extant early illustrations it would appear that some of the
original 'Stripper''s components were castings. Independent
confirmation of this was provided many years afterwards by pioneer
colonist James Umpherston,[63] when he recalled the destruction of
'much material of iron and timber *and castings*'[64] at Hindmarsh while
the first machine was being built. That being so, the castings must
have been produced by one (or possibly both) of the foundries in the
colony at the time — operated by Messrs Wyatt and Pybus
respectively. To this extent one at least of these two men also had a
small hand in the project.

In summary, then, it appears that the design evolved in Ridley's
mind somewhere between mid-1842 and July or August 1843. The

combined testimony of Ridley and Dunn indicates that Ridley based his ideas on his earlier study of the ancient Gallic harvesting method, and received no assistance from any other source whatever. Actual construction of a machine obviously began *before* any of the meetings organized by the 'Corn Exchange Committee' were held, but probably not very long before. The most likely time appears to be around late August or the beginning of September 1843. Ridley himself supervised the work, the whole of which was carried out in the workshops attached to his flour-mill at Hindmarsh. Publicity was kept to a minimum throughout the construction period. At least eight craftsmen were employed on the project in some capacity — John Dunn, John Dawkins, John Wilkinson, Samuel Marshall, and Messrs Cooke, Langman, Burton, and Isaac Hall. There may have been others as well — e.g., Thomas Munroe, who steered the first machine during the Wayville demonstration. John Godlee (a blacksmith) is another possibility. However in both these cases the evidence is too nebulous to be of much use.

According to John Dunn several days were taken up in making the replacement iron comb, and no doubt several more days were occupied before that with the unsuccessful initial trials carried out with the wooden comb. Thus it is clear that the machine in its original form (i.e., with wooden comb) must have been completed at least a week or so before the Wayville event. Indeed, certain press reports in late October and early November indicate that it was finished by the last week of October at the latest — a conclusion which has considerable significance in relation to J.W. Bull's later dispute with Ridley over priority of invention. A close examination of that aspect, however, must wait until another time.

The Structure and Mechanism of the Original 'Stripper'

Apart from the notices in the Adelaide newspapers, the earliest extant account of Ridley's original machine is that produced by the Colonial Engineer, Sergeant-Major Gardiner, in the first weeks of 1845. This arose from a request by Governor Grey for an official report on its construction and functioning — presumably for transmission to the colonial authorities in London. Gardiner's report concentrated mostly on the field performance of the machine, but part of his text also referred to drawings which were evidently attached. Unfortunately these are now lost, and only the verbal portion remains.[65] The earliest illustrations still existing (apart from the sketch by 'N.R.F.' reproduced previously as Fig. 1) are those in F.S. Dutton's book, *South Australia and its Mines* (1846), now reproduced below as Figs. 5 and 6. However, as will be shown in a moment, these are not an entirely faithful representation of the machine actually produced by Ridley.

As a preliminary to describing the structure and mechanism in detail, Dutton in his book first commented upon the immense value of the machine to the present and future development of the colony's wheat industry. He also remarked pointedly how well it was suited to local climatic conditions, noting that in the dry South Australian crops 'the corn separates from the chaff at the first blow of the beater'. Then

Figure 5. A woodcut illustration of the 'Stripper' in action. [Reproduced from F.S. Dutton, *South Australia and its Mines* (1846).]

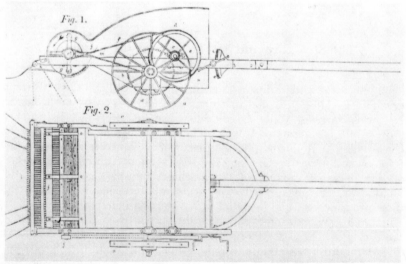

Figure 6. Diagram of the 'Stripper''s structure and mechanism. [Reproduced from F.S. Dutton, *South Australia and its Mines* (1846).]

followed some complimentary remarks on the efficiency with which it gathered the grain, with emphasis upon the extraordinarily small amount of grain wasted by the machine compared to hand-reaping. Dutton's illustrations and accompanying specification were as follows:

DESCRIPTION OF REAPING MACHINE.
Fig. 1. Side Elevation.   Fig. 2. Plan.
The letters correspond to both.

This machine is driven by two horses ... the carrying wheels aaa, are 4 feet in diameter, that on the off side is fixed to the axle, whilst the near wheel works in a box the same as an ordinary carriage wheel. To the inside of the off, or driving wheel, is attached a toothed rigger b, 30 inches diameter; this gears into the pinion c, on the shaft d, and gives motion to the fly wheel e, round which a cross belt f, passes, communicating with the pulley gg; this gives motion to the beaters hh, which makes 30 revolutions to one of the driving wheel; now the driving wheel, at a moderate horse walk, revolves 20 times per minute, giving to the beaters a velocity, $30 \times 20 = 600$ revolutions per minute, in the direction of the arrows.

At the fore end of the machine are six prongs, three on each side, embracing the entire width of the wheel track, and serving to collect the ears into the narrower range of teeth i, these extend into the cylinder, in the form of a comb, and, between them, the neck of the straw passes to j, (as shewn by the dotted lines 1, 2, 3) when, coming in contact with the beaters, the corn is struck out and thrown up the curve m, over which it falls into the body of the cart k.

The machine is propelled by a pole from behind, supported by two small wheels. The fore end of the machine is raised or depressed by turning the handle n, on the shaft of which is a pinion working in the segment rack 1. This arrangement enables the workman to adapt the machine to long or short straw. In the vignette [Fig. 5] ... the end of the cylinder is left open purposely to shew the beaters inside.[66]

A curious feature of Dutton's illustrations is that no 'chimney' (or even an opening) at the top for the escape of chaff and dust is shown. However some such 'chimney' was undoubtedly provided in Ridley's original machine (see, for example, Fig. 1), and indeed is mentioned by Dutton in his own written text:

[there was] ... a sort of chimney at the upper and back end of the large receiving box [through which] ... the greatest quantity of the chaff, makes its escape by the draught caused by the revolving of the beaters.[67]

Clearly, the omission of the chimney in Dutton's illustrations was unintentional — a careless oversight. However his verbal reference to the *greatest quantity* of the chaff being expelled through the chimney is illuminating, since it indicates how the earlier newspaper confusion over winnowing[68] may have arisen. Plainly, *some* winnowing action resulted from the draught caused by the beaters, but evidently not enough to fully cleanse the grain. Also significant was Dutton's further comment that Ridley hoped 'to add the perfect winnowing action' to his 'Stripper' design at some later time.[69] As it happened, Ridley never succeeded in doing so.[70]

From an inspection of Dutton's illustrations and those of others, it would appear that the main frame, the rear guiding pole, and possibly the sides and bottom of the receiving bin were made of wood. The blades of the revolving beater assembly were also flat pieces of wood. Illustrations such as Figs. 1 and 5 leave the impression that the large driving wheels may have been made entirely of iron, but this seems very unlikely at this early stage. Some other illustrations purporting to represent 'Strippers' of this period suggest wooden wheels with perhaps an iron tyre on the outside. Ridley's employment of Mr Cooke, the wheelwright (see earlier), also appears to indicate wooden wheels for the first machine, since otherwise it is difficult to account for Cooke being involved in the construction programme at all. The final comb was made from forged iron by Dunn and Dawkins, as already described, and replaced a wooden one which had proved unsatisfactory in the first instance.

Some of the other components were undoubtedly iron castings, produced in one or other of the two Adelaide foundries of the time. Likely items in this group are the flywheel, small belt pulley, gears, and the height-adjustment sector on the fore end of the guiding pole. To have made these items other than by casting would have been tedious and unnecessarily expensive, and especially so for the gears and the toothed sector. The various shafts and axles in the machine were iron forgings, and were probably made by the blacksmith, Mr Langman, in Ridley's own workshop. The same was probably true of the six frontal 'prongs' (for guiding the stalks into the comb) and a number of other small items, such as some of the harness fixtures attached to the guiding pole.

The shape of the cowl under the beaters and also that of the top cover for the body, suggest that both would most easily have been made up from sheet iron[71] — perhaps several small sheets riveted together — *if* such was procurable. It is known that sheet-metal was used for these items later in the century, but it is not clear whether suitable iron sheets would have been available in South Australia as early as 1843. It is possible, therefore, that for the first machines some other material may have been used such as, for example, canvas stretched over a light wooden frame. Fig. 6 does not indicate any such frame, but in a schematic drawing that kind of detail is not usually included in any case.

However Fig. 6 does show clearly the mechanical system employed to drive the beaters. Two speed-increasing elements were used; the first was a pair of ordinary toothed gears, and the other a crossed belt connecting two pulleys of unequal size. The belt was of course made of leather. According to F.S. Dutton,[72] this arrangement produced thirty revolutions of the beaters for every one of the ground-wheels, or a speed of about 600 revs. per minute at the ordinary walking pace of a horse. In practice this choice of ratios and speeds proved to be almost ideal, enabling the machine to harvest very cleanly but without damaging the grains by striking them so hard as to 'crack' them.

None of the reports describing Ridley's original machine gave a dimension for the breadth of the comb, although one in the press at the end of 1844[73] spoke of the framework on which the comb was mounted as being 'of the width of two or three feet'. On the other hand, if Dutton's diagram (Fig. 6) was drawn reasonably in proportion, then the breadth of the comb was evidently some 25% greater than the diameter of the ground-wheels, which Dutton stated were four feet in diameter.[74] Considering also that the machine was propelled by two horses, five feet or so for the breadth of the comb seems an entirely reasonable figure.

Operation of the 'Stripper' in the Field

A simple calculation,[75] based on the data provided in Dutton's specification of the 'Stripper' (reproduced earlier), shows that a blade of the revolving beater struck the ears of the growing crop *at about every* $1\frac{1}{4}$ *to* $1\frac{1}{2}$ *inches* of the machine's progress. This provides at once a ready explanation of its ability to harvest so very cleanly, even though it also gathered the grain quite rapidly. Each ear of wheat was struck not once, but several times, thus ensuring the removal of virtually every grain. Dutton himself made a point of noting that very little grain was wasted, but stated as well that one of these machines could fill a three-bushel bag (i.e., approx. 180 lb weight) every ten minutes. If a winnowing machine also happened to be on hand, he said, the wheat could be on its way to the mill within an hour of growing in the field.[76]

All the contemporary reports agreed that the machine required either two horses or two bullocks to drive it through the crop, pushing from behind to avoid trampling the unharvested area. Horses were preferred because of their faster rate of working and ease of handling. They were harnessed one on either side of the long guiding pole which, according to the illustration reproduced in Fig. 5, was itself supported on a separate pair of small wheels at the rear end. Adjustment of the height of the comb to suit the particular crop was obtained by choosing a suitable setting of the clamp-and-sector just behind the body of the machine (see also Fig. 6). However, because the effort required from the animals was both fairly considerable and sustained, it was found necessary in practice to replace them with a fresh pair after each hour or so of continuous working. Thus four, six, or even eight horses were customarily used, working in pairs sequentially.

According to Sergeant-Major Gardiner's official report (previously cited)[77] only two men were required per machine:[78]

> one to drive and steer, the other to assist in changing the horses.
>
> This latter man, in addition to the assistance thus rendered, collects the grain emptied from the machine on a tarpaulin and winnows it on the ground.

Note, incidentally, that Gardiner referred specifically to winnowing the grain *outside* the machine, showing once again that the earlier claims in the press concerning winnowing *by* the machine were incorrect.

At the same time it should be noted that the partial winnowing action provided inside the machine by the draught from the beaters and the 'chimney' gave rise to one of the rare criticisms ever made of Ridley's 'Stripper'. According to a recorded minute of the 'South Australian Agricultural and Horticultural Society'[79] this strong draught of air caused the seeds of a weed known as 'drake'[80] to be expelled with the chaff and dust, and thus to be spread across the field. These seeds then germinated the following year and grew more profusely than before, eventually causing the ruin of future crops. However Major T.S. O'Halloran (a farmer and prominent local citizen) suggested a remedy in a report which he wrote for the Colonial Secretary early in 1846. The problem with 'drake', he said, could be

> easily obviated by sowing clean seed upon clean land; or by allowing a dirty field to remain fallow for a season; the farmer's interests will soon oblige him carefully to attend to these points.[81]

Undoubtedly Ridley's 'Stripper' was at its best working in fully ripe wheat crops when the atmosphere was very dry. In damp or even humid conditions the machine was considerably less effective, and farmers soon learned to suspend harvesting operations on wet or cloudy days. Even in fine weather they found it worthwhile to avoid harvesting in the early part of the mornings, and to wait until the overnight dew had thoroughly evaporated. In the presence of any substantial amount of dampness the machine tended to choke the comb, and pull the stalks of wheat out of the ground instead of beating the grain off cleanly. Captain C.H. Bagot, another pioneer South Australian farmer, recorded that it was seldom his practice to work his 'Stripper' before 11.00 a.m. Before that hour, he said, the straw was inclined to be tough, and the threshing 'not so perfect' as later in the day.[82]

Used correctly, the Ridley 'Stripper' seems to have been a near-faultless machine from the farmers' point of view. Not only did it perform its task very efficiently, but also it appears to have been free of the kind of irritating mechanical and maintenance problems which plagued the American reapers during their early years of development.[83] The absence of cutters in Ridley's machine (which, as earlier British and American experience showed, would have required regular sharpening and adjustment) was a major factor in this respect. But in addition Ridley's design possessed another important virtue, mechanical *simplicity*. This, combined with a rugged construction, made it an implement well suited to the rough farming conditions applying throughout the colony during the early years.

John Ridley's Second 'Stripper'

Earlier in the discussion I conjectured that the commencement of construction of the original 'Stripper' was about the beginning of September 1843, or perhaps a week or two before that. Also I noted that, according to John Dunn's later evidence, all construction work (except for the replacement of the initial wooden comb) was completed well before the Wayville demonstration on 14 November. Thus the actual period required to build the first machine can be placed with reasonable confidence at somewhere between nine and twelve weeks.

Construction of a second machine was begun at Ridley's Hindmarsh workshops immediately after the dramatic success of the first at Wayville. This new machine is known to have been completed in eight-and-a-half weeks, or possibly even a little less — a fact which tends to confirm the estimate above of nine to twelve weeks for the first.

John Dunn afterwards described how, following the Wayville debut of machine No. 1, he himself was kept fully occupied for several weeks operating it on various farms in the Adelaide area. His published account of his experiences[84] during that time makes very interesting reading indeed. Ridley went ahead with the construction of his second machine without delay despite Dunn's absence from the workshop, since he felt sure that another would be needed during the course of that harvest if the whole crop was to be gathered. His judgement on this latter score proved to be correct, as did his belief that his other workmen at Hindmarsh could manage by themselves. Presumably either John Dawkins or John Wilkinson substituted for Dunn as 'leading hand' for the new project. Dunn later confirmed that he was absent from the workshop for the whole construction period of this second model, when he wrote:

> The first machine was such a success that all hands who could use a tool were set to work to build No. 2. I steered No. 1 all the time that No. 2 was being got ready.[85]

On 13 January 1844 a press report[86] announced that 'machine No. 2 was turned out of hand yesterday', and was being sent at once to

Mount Barker to take off some of the heavy crops there. The report also noted that machine No. 1 was 'undergoing some trifling repairs' before being dispatched to Mount Barker as well. By this time the gathering of all the crops on the Adelaide plains was virtually complete, and so the machines were no longer needed there. In the elevated Mount Barker district on the other hand the harvest was just beginning, because the wheat needed longer to ripen in the cooler conditions.

The repairs referred to were evidently quickly completed, and indeed there was never any suggestion that machine No. 1 had developed trouble of a serious nature. Simple routine maintenance appears to have been all that it required. In the interim the versatile John Dunn was allotted the rather tricky job of taking machine No. 2 up the steep mountain slopes to Mount Barker, and afterwards he acted as its driver and general overseer when it was put to work. In fact Dunn operated the new machine continuously on properties in the district until he was eventually recalled by Ridley, and another man (not identified in the reports) was detailed to take his place.

The Contribution Made by Ridley's First Two Machines at the 1843–44 Harvest

One of the more remarkable features of Ridley's original pair of machines was the rapidity with which they went into regular service in the field. Indeed it could be said that the very first machine did so on the day of its initial trial at Wayville, when it worked steadily for the whole afternoon and then continued in the same fashion over the following six days. According to F.S. Dutton, over those seven days all the grain from the entire Wayville crop of seventy acres was successfully collected.[87] Shortly afterwards it harvested seventy-two acres for a Mr George Emmett at Lyndoch Valley, and then a further eleven acres for one Thomas Whinnerah.[88] In fact the numerous reports indicate that, after Wayville, machine No. 1 was kept in constant work on the Adelaide plains until mid-January, when it was sent to join machine No. 2 at Mount Barker.

In the light crops around Adelaide the rate of working of machine No. 1 seems to have been consistently about ten acres per day. This was the figure quoted by F.S. Dutton, for example, and also by the Colonial Engineer in his official report,[89] as follows:

> According to ... Mr. Ridley's foreman [John Dunn?] ... the machine will reap and thrash about one acre per hour including stoppages, on an average perhaps about 200 bushels daily, the quantity depending of course upon the nature of the crop.'

In the latter part of January, and under the expert management of

John Dunn, machine No. 2 demonstrated that it could handle the much heavier crops around Mount Barker just as readily as machine No. 1 had dealt with the light ones on the coastal plain earlier. Within only a few days of its arrival in the district the new machine was reported in the press to be: 'cutting down[90] the tall corn of Mt. Barker at the rate of nine acres per day, and . . . giving great satisfaction.'[91]

Machine No. 1 performed similarly when it arrived after its minor repairs, and the two were kept constantly busy until the harvest was fully completed.

However some of the labourers of the district became somewhat apprehensive over the arrival of the new machines, fearing that they might deprive local men of work and so cause unemployment in their ranks. There was also an instance, apparently, where:

> a pious man named Horsell believed this invention was a wicked thing. On Sunday morning he knelt in the street opposite the machine and prayed God to send rain to spoil the crop and blast the machine — a thing of the Devil.[92]

By the time the harvest was over, however, the more enlightened members of the community had no doubts left concerning the value of the 'Stripper' for the future of the colony's wheat growing. Between them, Ridley's two machines had been able to harvest only a small fraction of the total acreage grown that year — perhaps not more than a thousand or fifteen hundred acres all told[93] — but the *manner* in which they had done so had been most impressive. The 'Strippers' had proved to be efficient, mechanically reliable, genuinely labour-saving, and above all, economically advantageous. Indeed, the use of such a machine provided the farmer with a multiple economy from the labour point of view, since it replaced the men formerly hired to thresh out the grain and those needed to collect and cart the stalks to the threshers, as well as most of those who had previously wielded scythes in the field. This meant that the level of production of wheat in the colony could now be made largely independent of the supply of harvest labourers. In future, production need be limited only by the availability of markets and the capacity of local manufacturers to supply sufficient machines to the growers.

The Manufacture of 'Strippers' During 1844

One other important aspect of the introduction of the 'Stripper' into South Australian agriculture remains to be noted, especially as it had a marked effect upon the subsequent production and use of the machines. This was that Ridley declined absolutely (on moral grounds) to seek a patent covering his invention. He wished it to be freely available, he said, to all who might wish at any time to make or use it. In short, he wanted it to be his gift to the colony.[94] As a result, no restrictions ever applied to the machine's manufacture from the very beginning. During the remainder of the nineteenth century many thousands of machines employing 'comb-and-beaters' were made in South Australia and elsewhere[95] by manufacturing firms and private individuals alike, and not a penny in royalties was paid to the inventor. Ridley himself was in fact the first to manufacture 'Strippers' commercially in South Australia (beginning in 1844), and this proved a very profitable undertaking for him as well as one of considerable importance to the colony. However, his success in this field was due to the high quality of the machines he produced, and not to any protection or advantage conferred under patent law.

Ridley's decision to set up in business manufacturing 'Strippers' for sale to farmers was nevertheless predictable in the circumstances. To begin with, a ready market was assured after the splendid performances at the harvest of 1843–44 of machines Nos. 1 and 2, already discussed. In addition, his team of craftsmen at Hindmarsh had by now acquired some degree of expertise in their construction. Indeed it is fair to say that Ridley's workshop was one of the few establishments in the colony suitably staffed and equipped at this time for such work. However it should also not be forgotten that, as a deeply religious man and (Wesleyan) lay preacher, Ridley almost certainly felt that he had a moral duty to use the technical resources he possessed for the benefit of the colony. Such a sense of obligation would have come naturally to a man of his particular outlook. But of course, if the venture happened to be virtually assured of financial success as well, so much the better.

Ridley began making machines for the local market almost as soon as the 1843–44 harvest season was over. Throughout 1844 his workmen at Hindmarsh were kept very busy trying to fill the orders which flowed in from farmers. The number of men actually engaged on this work does not seem to have been recorded, but by the end of November 1844 we find it reported by the *Southern Australian* that 'Mr. Ridley's machines [are] . . . now eight in number.'[96] In addition, taking advantage of the absence of patent restrictions on the design, several other colonial 'machinists' and even a few practically minded farmers had also begun constructing machines of the 'comb-and-beaters' type. Of these, the same report in the *Southern Australian* remarked:

> seven other reaping machines . . . are in course of construction, for parties chiefly resident on the plains to the southward. They all, we believe, differ in some respects from each other, and from Mr. Ridley's.

Apparently all but one of the machines produced by Ridley during 1844 were sold locally, and gave satisfaction to their purchasers. (The odd one out went to Western Australia, to a farmer named Irwin.)[97] One of Ridley's customers, Mr Hiram Manfull, testified that, with the aid of one man and four horses only, his new 'Ridley' frequently threshed ten acres per day.[98] Early in 1845 Manfull found that even better performance was obtained after he had modified the drive to increase the drum speed (or rotational speed of the beaters). The improvement, he found, was that harvesting in heavy crops was easier and the grain was noticably cleaner.[99] A fairly obvious explanation for the latter at least is that the greater speed of the beaters gave an enhanced draught of air, which in turn got rid of 'drake' seeds and other light rubbish more thoroughly.

At the 1844–45 harvest Ridley considerably expanded another activity he had begun the previous year, namely hiring out his machines (with drivers) to farmers who could not afford the capital outlay for a machine of their own. Once again, his motive was very likely a mixture of public-spirited concern and a shrewd recognition of the prospect of good profits to be made. By setting a moderate rate of five shillings per acre harvested,[100] Ridley soon attracted considerable custom. At that price, hiring a machine instead of men to reap by hand was a worthwhile proposition for the farmer, no matter what the size of the crop. Not only was the actual cost of harvesting well below that for hand reaping and separate threshing, but in addition the crop was gathered quickly with a minimum of waste. Moreover the farmer was relieved of the worry of finding suitably experienced (and sober!) labourers to cut and thresh his crop.

For Ridley the scheme was an equally good proposition. By harvesting up to ten acres in a day, one machine earned considerably more in hire fees than was paid out in wages to its two operators,[101] thus providing a generous return for the proprietor.

Significant Design Improvements Made During 1844

Ridley's machine No. 2 was undoubtedly similar in basic design to machine No. 1, but it was by no means identical. That this was the case is clear from the Adelaide *Observer*'s report when machine No. 2 first appeared in January 1844. It was, according to this, 'a decided improvement upon its predecessor both as to strength and efficiency'.[102] However the only differences between the two mentioned in any of John Dunn's published writings were (for machine No. 2):

> two low wheels and a short axle under the end of the [rear] pole, and the pole was fitted to the body with fitchels like a tip bullock cart.[103]

As noted earlier, Dunn worked both machines for considerable periods, and hence it is unlikely that he would have failed to mention other substantial modifications if such had been made. But at the same time Dunn's reference to small wheels under the pole of machine No. 2 as a new feature is puzzling, since it suggests that machine No. 1 was without such wheels. If this is the case, it reveals what must be a further error in the illustrations provided in 1846 by F.S. Dutton (see Figs. 5 and 6), and claimed by him to be representations of the original form of the machine.

Whether or not Ridley made any further changes in the design in machines made during the remainder of 1844 is uncertain. At the end of that year the *Southern Australian* newspaper reported what seemed to be a substantial improvement in the rate of working of his newer machines, but made no comment upon the possible reasons for it:

> We are informed that two of Mr. Ridley's machines lately thrashed fifteen acres of wheat each in a day. This was done at the Pine Forest, where the crops, we understand, average thirty bushels per acre, so that each of these machines would thrash 450 bushels of grain per day.[104]

It would be a mistake, however, to conclude from this that some kind of design improvement was necessarily responsible, since fifteen acres per day had also been claimed for machine No. 1 on at least one occasion at the previous harvest.[105] It thus seems that this higher working rate was due to other factors, such as particularly favourable crop conditions and (perhaps) superior skills on the part of some machine operators.

Important design improvements to the 'Stripper' were in fact introduced during 1844, but not by Ridley himself. The man responsible was Walter Paterson, who was farming at Mount Barker at the time, but who had originally been trained (in England) as a 'mechanic'. Ridley's machine No. 2 harvested a wheat crop for Mr Paterson, among others, while it was operating in the Mount Barker district in January 1844, and naturally Paterson took the opportunity of observing the machine closely during the time it worked on his property. Shortly afterwards Mr Paterson set about constructing a machine for his own use at future harvests, based on the 'stripping' principle and closely following Ridley's main structural arrangements, but also featuring three substantial improvements of his own devising.

The first was a change to the drive for the beaters, in which Paterson set out to replace the clumsy crossed-belt final drive of the Ridley design with one using a simple open belt. To do this, Paterson had first to cause the flywheel to rotate in the opposite direction to Ridley's (see also Fig. 4). This was accomplished by driving the flywheel from the ground-wheel through an internal gear, instead of using a straightforward external gear-drive as on the original 'Ridley'. Paterson's internal gear was a little more difficult to manufacture and thus more expensive, but at the same time its larger diameter (for the same shaft spacings) provided a welcome increase in the rotational

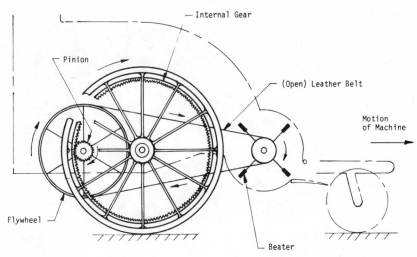

Figure 7. Schematic diagram of Walter Paterson's improved drive for the revolving beaters

speed of the beaters and in addition it allowed the open-belt configuration he desired for the final drive. Also, as Paterson had predicted, the open-belt arrangement proved to be more reliable in service and was moreover subject to less wear and a lower friction loss.

Paterson's second improvement consisted in arranging for the horses to *pull* the machine from the near-side front corner[106] instead of pushing it from the rear. This produced several benefits. The elimination of the long rear pole (and the extra set of small wheels necessary to support it) simplified the machine's construction and reduced its overall weight. The task of steering was also made much easier, because the driver was now positioned further forward and was thus able to watch the line taken by the horses more closely. The machine's handling characteristics — and especially its manoeuvrability — were likewise markedly improved by the reductions in length and weight. And finally, since a lighter machine required less effort from the horses to draw it, less frequent stoppages for fresh animals were necessary.

Paterson's third improvement was a simple but nonetheless important one: the provision of a *seat* for the driver on the front near-side of the machine, about level with the beaters. Although apparently a trifling thing, this gave the driver much needed stability on the moving machine, and hence improved control over its line through the crop. John Dunn was one who especially noted the value of such a seat for the man driving when (more than forty years afterwards) he recalled the discomfort and difficulty he had experienced while 'perched ... on the body' of Ridley's No. 2 machine during the 1843–44 harvest.[107]

Figure 8. A prize-winning 'Stripper' manufactured in the 1870s, featuring Paterson's modifications. [Reproduced from W. Harcus (ed.), *South Australia* (1976), p.359.]

All three of these modifications introduced by Paterson were decided practical improvements upon Ridley's original design, and contributed greatly to a still higher operational efficiency in the field. Walter Paterson himself reportedly earned between £400 and £500 each harvest season during the later 1840s, by using his machine to gather in crops for other farmers.[108] His design improvements were also eagerly adopted by John Ridley and other machine makers, and soon became part of the standard machine. In fact, Paterson's new features were included in all new machines made after about mid-1844, right up until the end of the nineteenth century and beyond. All three are clearly evident in the machine illustrated above, for example, which was produced some thirty years afterwards and is representative of the best machines then available.

Rapid Growth in the Popularity of the 'Stripper': 1844–53

Predictably, the success of Ridley's machine in its first year or two of operation had a marked influence upon the subsequent development of South Australian wheat-growing. The settlers quickly recognized the benefits of harvesting by the new method, and it became the goal of virtually every wheat farmer in the region to own at least one of these useful machines. The resulting demand was responsible for the rise of a considerable number of agricultural implement manufacturing firms in and around Adelaide during the later 1840s and the 1850s. Some of these developed into large and flourishing establishments, and not a few continued to prosper well into the twentieth century. Indeed several of them went on to acquire international reputations, selling their products widely in countries overseas as well as throughout the whole of Australia.[109]

From 1844 onwards the farmers' earlier hesitation to expand their areas under crops was replaced by a general mood of optimism. Huge increases in the acreages sown to wheat occurred year by year, and continued virtually unchecked for several decades. In fact, so dramatic was the progress of wheat culture in this period that, a mere thirty-one years after the first appearance of the 'Stripper', Captain C.H. Bagot could quite truthfully assert that:

> the excess of grain has been, on an average, equal to seven years consumption of the people . . . who all consume the best white bread without stint . . . for each year, a fact for which no paralel [sic!] can be found in the history of the world.[110]

Well before the colony celebrated its fiftieth anniversary (in 1886) there was no doubt on the part of any South Australian resident that 'its [the 'Stripper''s] adoption in our cornfields [has] placed South Australia at the head of the colonies as a producer of wheat'.[111]

Of course, it was the amazing *economy* of Ridley's machine which, as much as anything, made this extraordinary development in wheat growing possible. Published references to this particular feature were made by Captain Bagot and other farmers after they had used the machine on their own properties for only a season or two. During the 1844–45 harvest season, for example, Bagot claimed that he had been able to gather his wheat and prepare it for market (i.e., including winnowing and presumably bagging) at a cost of only $3\frac{1}{2}$ pence per bushel.[112] Later it was shown that Bagot probably underestimated some of his production costs on this occasion,[113] but still his basic point (that the 'Stripper' permitted very cheap harvesting indeed) was never disputed.

The following year, and using a much larger machine than Bagot's, another local farmer named Major T.S. O'Halloran calculated that his total harvesting costs for a crop of about 190 acres had been 9/8½ per acre.[114] In this case there seems every reason to trust the figures, since O'Halloran detailed his expenditure very carefully. He even included, for example, 'sixty-four gallons of Cape Wine' consumed by the hands he employed. Assuming that his crop was average for the Adelaide plains region (say, 20–30 bushels per acre), Major O'Halloran's figures represented costs of between 4d. and 6d. per bushel of grain up to the time it was delivered to the miller for grinding. Also, according to

O'Halloran, gathering the harvest by hand would have taken three times as long, and would have cost at least three times as much. And further, the loss of grain due to shedding, vermin damage and rain would undoubtedly have been far greater without the use of the machine. The South Australian community, he said, owed John Ridley a very deep debt of gratitude.

Comparing O'Halloran's estimate of 9/8½ per acre with the total costs involved in harvesting by hand, it may be shown that his claim for a one-to-three ratio was entirely realistic for the time. At the 1842–43 season, for instance, farmers commonly paid wages of 15/- and more per acre for hand-reaping alone, to which must be added (for a valid comparison) further costs for threshing, winnowing, and bagging the wheat before it could be sent off for milling. In addition, a considerable capital outlay was required beforehand for threshing machines, 'stackyards', and barns in which to winnow and store the grain after it was cut by the scythe-men. When the crop was machine-harvested all these were unnecessary, since all the operations could then be carried out in the harvest field itself and *on the same day*.

As might be expected, the year 1845 saw greatly increased activity in 'Stripper' construction. In mid-November it was reported in the press[115] that Ridley had almost completed a further seven machines, and that other makers had also been busy. Six of Ridley's new machines were about the same size as his previous models (but lighter and stronger), while the seventh was described as:

> a large double acting one, with two sets of beaters, and intended to be drawn by oxen. It is nearly seven feet in width, and both wheels are geared to work the beaters.[116]

Apparently this was the large machine purchased by Major T.S. O'Halloran and used by him with such success at the following harvest, as just described.

According to remarks printed in the *South Australian Register*, the total number of machines available for use at the 1845–46 harvest was expected to be 'no fewer than fifty [*more*]' than for the previous season, manufactured by 'Mr. Ridley and competitors in town and country'.[117] Two days later, however, the *South Australian* gave a more cautious estimate of 'upwards of a dozen' new machines that year, bringing the total number in the colony to 'upwards of twenty'.[118] Also according to this report, six of the new machines had been manufactured by Ridley. Four of the others are known to have been made by James Adamson,[119] whose business later developed into one of the largest and most famous implement manufacturing firms in South Australia.

By the end of the 1845–46 harvest it was clear that the traditional reaping of wheat crops using hand-tools was destined soon to disappear from the South Australian scene. Indeed, in his book published only a few months later F.S. Dutton went so far as to claim that:

by this time [1846] ... the greatest part of all the wheat grown in the colony is harvested by this [Ridley's] machine, causing an enormous saving of labour and expense.[120]

At first sight the first part of this assertion appears to be impossibly exaggerated. Almost 19,000 acres of wheat were harvested at the end of 1845,[121] but as already shown only some twenty or thirty 'Strippers' were then in existence. In addition it was necessary in South Australia to take off crops within a period of only two or three weeks, because of the tendency of the grain rapidly to over-ripen in the intense summer heat. Hence, assuming a rate of working for each of (say 25) machines of 10–15 acres per day, the 'greatest part' of the 1845 crop (say 10,000 acres at a minimum) would evidently have required $10,000/25 \times 12\frac{1}{2} = 32$ working days for all the machines in the colony.

However a resolution of this apparent dilemma was suggested two years later by the colonial author G.B. Wilkinson, when he pointed out that:

on the plains the wheat harvest commences in November, but in the hilly districts not until December or the beginning of January, according as the season is wet or cold, or fine and warm.[122]

In other words, although the harvest period for *individual* wheat crops was quite short, those in various regions of the colony could be harvested more or less sequentially over a season lasting more than two months. That being so, Dutton's claim that mechanization in wheat-harvesting was well advanced in South Australia by 1846 could have been, and indeed probably was, entirely correct. Dutton's further point — that large savings in labour and expense resulted from the use of the 'Stripper' — was also correct, and it was reinforced in G.B. Wilkinson's account (already cited) as follows:

Reaping by hand costs from 12s. to 15s. per acre; but if the crop be cut [?] by the reaping machine ... the cost is only about 8s. for the same quantity.[123]

Wilkinson did not state whether his figure for hand-reaping included the cost of threshing or not. However, even if it did, it is interesting to note that the capital outlay required for one of Ridley's machines (approximately £50) could be recovered in harvesting a mere 150–250 acres of crop at a single season.

According to the *South Australian* newspaper early in 1847[124] 'at least thirty' machines had been employed at the 1846–47 harvest season just ending, and the number continually increased over following years. An account of contemporary South Australian conditions published by an 'Old Colonist' in or about the year 1851[125] strongly suggested that by then the mechanisation of harvesting in the colony had reached quite an advanced stage, and perhaps was even approaching completeness. Disappointingly, this author did not attempt to estimate the actual number of 'Strippers' then in use, but at the same time he left no doubt in his readers' minds that hand-reaping had become a rarity. Describing a tour he had made through outlying farming areas (i.e., well away from Adelaide) during the 1850–51 harvest, 'Old Colonist' remarked several times that, on all sides, reaping and winnowing machines were 'everywhere on the lands'. For example, he said, at the height of the harvest in the Willunga district[126] these machines were to be found at work 'in almost every field'. Throughout his journeys, he further remarked, the only hand-reaping he saw was in very small crops, or in particularly remote areas, and this work now seemed to be mostly confined to aboriginal labourers. As a point of sociological interest, however, he also noted that the blacks were paid at the same rates as whites similarly employed.

Thus it is seen that there exists a clear difference of opinion on the advance of harvest mechanisation in the colony during these early years between some contemporary writers, like 'Old Colonist' and F.S. Dutton, and certain modern scholars such as the Australian economic historian Dr. Edgars Dunsdorfs. In his *The Australian Wheat Growing Industry: 1788–1948* (1956) Dunsdorfs claimed that mechanisation in South Australian wheat harvesting was still far from complete well into the 1860s, but he presented no evidence to support this contention. In contrast to this we note that 'Old Colonist' and Dutton both strongly asserted that the major part of the colony's grain was gathered by machines by 1850 or earlier. If their accounts are correct (and it should be borne in mind that both were *eye-witness* accounts), then they have an important implication which seems to have been entirely overlooked by historians of agricultural technology everywhere. This is that, sometime between 1846 and 1851, South Australia achieved the mechanisation of the majority of its wheat harvesting, and was thus *the first region in the world to do so*. As I have shown elsewhere[127] workable grain harvesting machines were developed both in Britain and America much earlier than in South Australia (though employing a quite different operating principle), but their introduction on any significant scale was delayed in both regions because (in Britain) the social circumstances were against them and (in America) mechanical reliability was lacking. Indeed, the widespread adoption of machines in both those places did not begin until about the mid-1850s.

Later Proliferation of the 'Stripper', and Concluding Remarks

As far as is known John Ridley himself continued to manufacture and sell 'Strippers' until shortly before he left the colony (in March 1853) to live in England. At that time Ridley was in fact still only in his late forties, but already he had amassed a comfortable fortune from his farming, flour-milling, and manufacturing activities, and his speculations in mining and in land. As a result he was in a position to retire early and return to England as a gentleman of means. The entire colony was saddened by his departure, for apart from having invented and introduced the 'Stripper' at precisely the right time in the first instance, Ridley had also provided much-needed technical and business leadership in the manufacture of the machines for almost a decade afterwards. As was indicated earlier, Ridley's manufacturing activities at Hindmarsh were both vigorous and profitable, but the profits he made were earned in open competition with all others who chose to make 'Strippers' for sale.

However, important as the machine had been to the colony during Ridley's time there, it was to prove equally if not more valuable in the years after he had left. It was, for example, a vital element in the survival of South Australian wheat production after 1851, when thousands of farm labourers defected to the rich goldfields discovered in that year in the neighbouring colony of Victoria.[128] Despite the loss of some 15,000 men during 1851 alone,[129] South Australia fared the best of all the Australian colonies, experiencing only a slight stagnation in grain exports between 1851 and 1855. Certainly the supply of labour on the farms was drastically reduced and the hire of farm workers became very much more expensive,[130] but at no time was there a serious possibility of a collapse in the colony's wheat industry. What actually happened was that:

> the gold rush caused the price [of wheat and flour] to skyrocket and these high prices coincided with good harvests. A boom resulted and the great proceeds were immediately invested in land *and machinery*.[131] Consequently a rapid expansion of the wheat-growing area followed.[132]

It should be understood, however, that here the term 'machinery' referred not only to the 'Stripper' (which indeed was virtually the exclusive choice of harvesting implement in the region), but also to implements for use in such operations as land clearing, ploughing, sowing, etc.

In the event, the progress of the local wheat industry was disturbed very little, and this is reflected in the figures in Table 2 following. For instance, the colony's population approximately doubled in the decade 1850–1860, but the total acreage under wheat increased six-and-a-half-fold and earnings from exports rose by a factor of more than thirteen. In the following decade the area sown to wheat again more than doubled, and then it almost tripled once more between 1870 and 1880, accompanied by an approximately five-fold increase in revenue from the export of grain and flour. The peak of this development was eventually reached about the year 1884, when almost

TABLE 2[133] South Australian Wheat Statistics: 1850–1884

Year	Population	Wheat acreage	Av. yield (bush./acre)	Total production (bushels)	Wheat and flour exports (bushels)	Value of breadstuffs exported
1850	63,700	41,807	16.0	668,912	186,000	£38,312
1852	68,663	65,877	10.5	691,708	442,000	£212,566
1854	92,543	89,945	8.0	719,560	429,000	£316,217
1856	104,708	162,012	25.0	4,050,300	1,177,000	£556,371
1858	118,340	188,703	11.2	2,113,474	1,428,000	£525,398
1860	124,112	273,672	13.1	3,585,103	1,286,000	£499,102
1862	135,329	320,160	12.0	3,841,920	2,469,000	£633,241
1866	163,452	457,628	14.3	6,544,080	1,868,000	£645,401
1870	183,797	604,761	11.5	6,954,752	1,644,000	£470,828
1873	198,075	784,784	7.9	6,199,794	6,685,000	£1,711,746
1875	210,442	898,820	12.0	10,785,840	7,634,000	£1,680,996
1877	236,864	1,163,646	7.8	9,076,439	3,703,000	£1,184,368
1880	267,573	1,733,542	5.0	8,667,710	11,029,000	£2,469,720
1882	293,509	1,746,531	4.2	7,335,430	5,961,000	£1,551,106
1884	312,781	1,942,453	7.5	14,568,398	12,593,000	£2,185,057

two million acres of wheat were harvested and export earnings from grain and grain products exceeded £2 million sterling.

At that stage almost all the land in South Australia available and suitable for growing wheat had been taken up, and the regional centre of growth in the industry had shifted across the border to Victoria.

However, as far as South Australia was concerned, there is another point which should also be made concerning the earlier stages of development. It is that, although the increases in wheat acreages from the late 1840s onwards were extraordinarily rapid, the growth in the manufacture and use of the 'Stripper' in the same period was even more so — due, of course, to the ever-dwindling supply of farm labour. Unfortunately no reliable record of the numbers of the machines actually made in the 1850s and afterwards is available, but an estimate can be made from press reports and other documentary sources.

At least 200 machines are believed to have been employed at the 1850–51 harvest, and it is certain that the number increased rapidly in following years. For instance, there is firm evidence that 350 *additional* 'Strippers' were produced by South Australian makers in the year 1856 alone.[134] At the 1860–61 harvest season, it has been estimated, between 1,600 and 2,000 machines were in the field, and the number then approximately trebled in the next ten years. The number of agricultural implement manufacturing *firms* in the colony also increased quite dramatically. In a book published in 1866, pioneer colonist Anthony Forster noted that there were then already forty-two such firms in existence,[135] while from William Harcus's work of a decade later we learn that the number had by then grown to eighty-six. Also, according to Harcus, the production of these firms was still 'chiefly . . . reaping and winnowing machines'.[136]

It has been estimated that, when South Australian wheat-growing levelled off around 1884, a total of between 16,000 and 20,000 'Strippers' were in use in the colony and several thousands more in the drier (Mallee) region in neighbouring Victoria. However it is not the purpose of this paper to deal with the progress of the South Australian wheat industry during this later period. Neither will the associated topics of the rise of tractor power and the eventual replacement of the 'Stripper' by the modern combine-harvester be pursued in any way. Our chief concern has been, and remains, the 'Stripper' itself and its effect upon the early development of wheat culture in its region. Nevertheless, two further observations upon the machines of the 1870s and early 1880s might perhaps be offered in conclusion as a means of making wider points concerning the machine. The first is that in his book of 1876 (previously cited) W. Harcus noted particularly that the usual duty of a 'Stripper' at that time was about seven or eight acres a day[137] — that is, about the same as was reported over thirty years before. From this it follows that the machines of Harcus's day can have been little if at all larger than those manufactured by Ridley himself. The second stems from a statement made by pioneer colonist Captain C.H. Bagot in a letter to his old friend John Ridley at the beginning of 1875, as follows:

> although several improvements have been made in the construction of the implement [the 'Stripper'] there have not been any in principle. It is still exactly as you invented it, the comb, the beating drum driven by the wheel revolving as it passes over the land, and the body of the machine for receiving the stripped grain. All these are unaltered in principle.[138]

In other words, apart from the practical modifications introduced by Walter Paterson in 1844 (previously noted), and minor improvements for strengthening or reducing the weight of individual elements, the machines of the 1870s were much the same as those of the early years.

Thus we have in the South Australian 'Stripper' an invention which not only was extraordinarily successful in the beginning, but which also had a remarkably successful subsequent 'career' in terms of widespread adoption and stability of design. As shown earlier, the machine was an unqualified success at its first public showing at Wayville on 14 November 1843, and it evidently required no modifications of any sort before it went into regular daily service in the harvest field. Thereafter, as attested by Captain Bagot's and other accounts, both the basic form of the machine and its principle of operation remained absolutely unaltered for a period considerably in excess of thirty years. Indeed, the first major design change to the 'Stripper' was the addition of an internal winnowing mechanism, which was attempted (with partial success) by George Marshall in the mid-1870s,[139] and finally accomplished by Hugh Victor McKay in 1884.[140] History shows that this kind of record — i.e., immediate success upon introduction and long service thereafter without significant change of design — has been most unusual among new inventions of any kind and in any age. Certainly no comparable claim can be made for any other new cereal-harvesting machine introduced anywhere in the world either before the 'Stripper' or since.

Notes

1. L.J. Jones, 'The Early History of Mechanical Harvesting', *History of Technology*, Vol.4 (1979), 101–48.

2. It is perhaps opportune here to note a new contribution on this and related topics which has appeared since I prepared my own earlier article (n.1) — viz., G.R. Quick and W. Buchele, *The Grain Harvesters*, published in the U.S.A. by the American Society of Agricultural Engineers in late 1978. The treatment of the inventors of this early period and their machines is brief but well researched, and significant new information concerning one or two of them is included.

3. See pp.122–6 of my earlier article (n.1 above).

4. Ibid. See especially pp.101–3.

5. For example, it is still asserted in some current works that the 'Stripper' performed all three basic harvesting operations — viz., reaping, threshing and winnowing. In fact it performed the first two *only*, and winnowing remained a separate task until H.V. McKay introduced his new design in 1884.

6. Wakefield's principal writings are now available in a single volume: see M.F. Lloyd Pritchard (ed.), *The Collected Works of Edward Gibbon Wakefield* (1968).

7. At this time (the 1820s and early 1830s) unemployment and poverty among the working classes were serious problems in Britain. Among the principal causes were the rapid introduction of machines to perform tasks previously carried out by hand, and the difficulty of re-absorbing into the labour force the soldiers who had returned from the Napoleonic wars.

8. For Wakefield's views on these, see his *Letter From Sydney* (1829), and *England and America* (1833). Both are included in Pritchard (n.6 above).

9. They are well discussed, for example, in E. Hodder, *The History of South Australia* (1893).

10. These included *An Act for Regulating the Sale of Waste Lands in the Australian Colonies and New Zealand* (22 June 1842), and *An Act for the Better Government of South Australia* (15 July 1842), 5 and 6 Victoria c.36, and 5 and 6 Victoria c.61, respectively.

11. A few differences between South Australia and the other colonies were retained however. For instance, the Wakefield principle of promoting immigration from the proceeds of land sales was retained in a modified form, and the original provision barring the entry of convicted felons was left unchanged.

12. Note that no population figures can be given for 1841 and 1842 because no census was taken in either of those years.

13. This passage from one of Grey's private letters was quoted in J. Milne, *The Romance of a Pro-Consul* (1899), p.64.

14. E.g., the American machines of Hussey and McCormick, which had been introduced a decade previously.

15. Two diagrams and a detailed description of Bell's reaper were included in J.C. Loudon, *An Encyclopedia of Agriculture*, 2nd edition (1831), pp.422–7, and presumably in following editions.

16. 'Agricola' to the Editor, *South Australian Register* (26 August 1843), p.2, col.3.

17. *South Australian Register* (2 September 1843), p. 3, col.2.

18. It is difficult to be sure of the precise number, since the reports of the meetings do not always make clear whether certain individuals were contestants, representatives for others, or mere spectators.

19. This particular piece of information had in fact been given to the Committee nine days earlier by a friend of Ridley's (not identified in the reports) who had spoken on his (Ridley's) behalf in the latter's absence — see the report in the Adelaide *Observer* (16 September 1843), p.4, col.1.

20. Adelaide *Observer* (23 September 1843), p.6, col.2.

21. Ibid.

22. A report of the trial of Marshall's machine appeared in the Adelaide *Observer* (4 November 1843), p.5, col.2.

23. This was done in a letter to the press in which Bull also claimed for himself the invention of the principle of the machine. See J.W. Bull to the Editor, Adelaide *Observer* (15 March 1845), p.7, col.1.

24. This was the description of the machine employed in the report of the Adelaide *Observer* (18 November 1843), p.5, col.1.

25. Adelaide *Observer* (28 October 1843), p.5, col.2.

26. See, for example, the *South Australian Register* (1 November 1843), p.2, col.5.

27. See John Dunn's long letter in the *Mount Barker Courier* (11 June 1886). It was republished in the *South Australian Register* (19 June 1886), p.7, cols.5,6,7.

28. Adelaide *Observer* (4 November 1843), p.5, col.2.

29. *South Australian Register* (15 November 1843), p.2, col.3.

30. *Southern Australian* (17 November 1843), p.2, col.4.

31. Ibid.

32. Adelaide *Observer* (18 November 1843), p.5, col.1.

33. See n.30 above.

34. Ibid.

35. This was almost certainly Robert Gouger who played a leading part in organizing the setting up of the colony in the first place, and who subsequently became South Australia's first Colonial Secretary.

36. See F.S. Dutton, *South Australia and its Mines* (1846), p.210.

37. See n.23 above.

38. See Bull's letter in the *South Australian Register* (4 June 1886), p.3, col.8.

39. The award was recorded in *South Australian Parliamentary Debates* (5 September 1882), col.844.

40. His letter was written in London and dated 18 March 1886. It was published in the *South Australian Register* (6 May 1886), and again in the Adelaide *Observer* (8 May 1886).

41. Discussed in detail in my previous article (see n.1 above), in K.D. White, *Agricultural Implements of the Roman World* (1967), and elsewhere.

42. Quoted from an article by John Dunn entitled 'Memories of Eighty Years', in the *Mount Barker Courier* (17 December 1886), p.4. (Note that this was one of a series of articles, all identically titled, published by Dunn in this newspaper over a period of some months.)

43. John Dunn to the Editor, Adelaide *Observer* (20 December 1890), p.23, col.2.

44. See n.27 above.

45. See n.40 above.

46. See J.C. Loudon, *An Encyclopedia of Agriculture*, 1st edition (1825). Meikle's threshing machine was illustrated on p.127, while a description and conjectural diagram of the 'Vallus' was included on p.26.

47. Adelaide *Observer* (18 November 1843), p.5, col.1.

48. See n.36 above.

49. A. Forster, *South Australia: its Progress and Prosperity* (1866).

50. This was noted in a private letter written by J.G. Coulls (or Coutts?) and dated 21 March 1876, and quoted in turn in a letter by R.F.C. Mau published in the Adelaide *Advertiser* (31 August 1928).

51. See n.17 above.

52. Adelaide *Observer* (9 September 1843), p.4, col.3.

53. Quoted from a letter by Dunn published under the heading 'The Late Mr. Dawkins' in the Adelaide *Observer* (20 December 1890), p.23, col.2.

54. See above, n.27.

55. Note that this is further evidence that the basic construction of Ridley's first machine was completed well before the public demonstration at Wayville on 14 November. As such, it is pertinent to the dispute between Ridley and Bull over priority of invention.

56. *South Australian Register* (19 June 1886), p.7, col.6.

57. This is stated in an official pamphlet, entitled 'Honoured at Last', marking the unveiling of a marble bust of John Ridley at Roseworthy Agricultural College, South Australia, in 1913.

58. See the Rev. W. Gray's article entitled 'The History of the Invention of the Ridley Stripper', in *Agriculturalist and Review* (4 August 1933).

59. See A.T. Saunders to the Editor, Adelaide *Register* (21 January 1928), p.13, col.4.

60. See n.34 above, and text, pp.65–6.

61. Fred. Burton to the Editor, Adelaide *Advertiser* (18 August 1928), p.19, col.1.

62. Samuel Marshall to J.W. Bull, letter dated 24 November 1875 (MS., document No. 1076/1 in the Archives of the State Library of South Australia, Adelaide).

63. James Umpherston to the Editor, *South Australian Register* (26 May 1886), p.3, col.8.

64. My emphasis (L.J.J.).

65. The verbal portion of Gardiner's report is now preserved in the Archives of the State Library of South Australia, Adelaide. It is handwritten, dated 13 January 1845, and marked 'Colonial Secretary's Office Docket No. 42'.

66. F.S. Dutton, op. cit. (see n.36 above), pp.215–16.

67. Ibid., p.211.

68. See n.47 above and text, p.70.

69. F.S. Dutton, op. cit., p.211.

70. This was eventually accomplished by H.V. McKay of Kamarooka, Victoria, in 1884.

71. Sheet steel was of course unknown at this early time.

72. F.S. Dutton, op. cit., p.215.

73. *Southern Australian* (20 December 1844), p.2, cols.4,5.

74. F.S. Dutton, op. cit., p.215.

75. The calculation is as follows:

Number of blades in the beater = 4
Diameter of ground-wheels = 4 feet (48 inches)
Revs. of heater assembly per rev. of ground-wheel = 30
Horizontal distance travelled for each rev. of ground-wheel = $\pi \times 48$ inches

Hence,

$$\text{distance travelled per 'strike'} = \frac{\pi \times 48}{4 \times 30}$$
$$= \underline{1.26 \text{ inches}}$$

Alternatively,
Approx. walking speed of horses = 3 miles per hour
Rotative speed of beater assembly = 600 r.p.m.

Hence,

$$\text{distance travelled per 'strike'} = \frac{4 \times 5280 \times 12}{60 \times 600 \times 4}$$
$$= \underline{1.32 \text{ inches}}$$

(Note that the two calculated values agree with approx. one-sixteenth of an inch, or about 5%.)

76. F.S. Dutton, op. cit., p.210.

77. See n.65 above.

78. This is supported by Dutton's account (n.36), and also by numerous reports in the press in 1843 and afterwards.

79. For a meeting of the Society held on 25 January 1845.

80. 'Drake', or 'darnel', is a type of grass (Lolium temulentum) which is similar to rye grass and is commonly found in Australian wheat country. It produces seeds which are much smaller and lighter than grains of wheat.

81. Quoted from T.S. O'Halloran's report to the Colonial Secretary on the performance of his (unusually large) 'Stripper' which he nicknamed 'Paddy Whack'. (MS., dated 23 March 1846.) A copy is filed in the Archives of the State Library of South Australia, Adelaide (File No. 38).

82. See Bagot's letter in the *South Australian* (17 January 1845), p.3, cols.3,4.

83. Discussed in my earlier article (see n.1 above).

84. John Dunn to the Editor, *South Australian Register* (19 June 1886), p.7, cols.5,6,7.

85. Ibid.

86. Adelaide Observer (13 January 1844), p.5, col.2.

87. F.S. Dutton, op. cit., p.210.

88. See E.H. Combe, *History of Gawler: 1837–1908* (1910), p.352.

89. See n.65 above.

90. This, of course, cannot be taken literally, since the machine in fact had no cutters.

91. *Southern Australian* (23 January 1844), p.3, col.2.

92. Quoted from an article entitled 'The History of the Invention of the Ridley Stripper' by Rev. William Gray, in *Agriculturalist and Review* (4 August 1933).

93. This estimate was arrived at by assuming that the harvest period finished about the end of January, and so counting the maximum possible number of working days for the two machines, as follows:

Machine No. 1 (14.11.43 to 31.1.44) — 78 days
Machine No. 2 (13.1.44 to 31.1.44) — 18 days

TOTAL — 96 days

Rate of working (per machine) = 10–15 acres/day

Hence:

Total area harvested = 960–1, 440 acres (approx.)

94. This was widely reported at the time and afterwards. It was officially recorded, for instance, in a resolution of thanks to Ridley passed by the South Australian Agricultural and Horticultural Society on 10 February 1853, and noted in the Society's *Minutes*. (The reason for the resolution was that Ridley was shortly to leave the colony for retirement in England.)

95. For example, in the adjacent colony of Victoria.

96. *Southern Australian* (29 November 1844), p.3, col.4.

97. Initially (Major) Irwin experienced some difficulty with his machine and threatened to return it as unsatisfactory. However, after advice from Ridley that he should not use it in damp conditions, no further complaint was made.

98. *Southern Australian* (20 December 1844), p.2, col.4.

99. See also *South Australian* (10 January 1845), p.3, col.2.

100. This was noted in the *South Australian Register* (16 December 1843), p.2, col.5, in a sub-editorial article devoted to Ridley's machine.

101. According to the *Statistical Year Book* for the colony for 1845, this would have been about 3/6 per day for each man.

102. Adelaide *Observer* (13 January 1844), p.5, col.2.

103. Quoted from John Dunn's published letter in the *Mount Barker Courier* (11 June 1886).

104. *Southern Australian* (20 December 1844), p.2, cols.4,5.

105. See Adelaide *Observer* (16 December 1843), p.5, col.1.

106. i.e., 'side-draught'. Note, however, that this feature had appeared more than twenty years earlier in some of the English 'cutting' reapers (see n.1 above).

107. See n.103 above.

108. This is according to Rev. William Gray (see n.92 above).

109. The firms of James Martin and Co. and Horwood-Bagshaw are two examples.

110. C.H. Bagot to John Ridley (MS., dated 28 January 1875). The original letter is now filed in the State Library of South Australia, Adelaide, as document No. 1053/29.

111. Quoted from an article entitled 'Ridley's Reaping Machine', in the Adelaide *Observer* (8 May 1886), p.9, cols.3,4.

112. C.H. Bagot to the Editor, *South Australian* (17 January 1845), p.3, cols.3,4.

113. See editorial comment on Bagot's letter (n.112) printed in the same issue of the *South Australian*.

114. See T.S. O'Halloran's report to the Colonial Secretary, previously cited (n.81).

115. *South Australian Register* (19 November 1845), p.2, col.4.

116. Ibid.

117. Ibid.

118. *South Australian* (21 November 1845), p.3, col.3.

119. This was stated in an advertisement placed by Adamson in the *South Australian* (9 December 1845), p.2, col.2.

120. F.S. Dutton, *South Australia and its Mines* (1846), p.210. Note, however, the striking contrast between this statement by Dutton and that of E.J.T. Collins describing the (much later) situation in Britain. According to Collins, 'by 1871 perhaps 30 per cent of the British corn area was cut by machine, rising to about 80 per cent in 1900' (E.J.T. Collins, *Sickle to Combine* (1968), p.6).

121. See Table 1.

122. G.B. Wilkinson, *South Australia; its Advantages and Resources* (1848), p.68.

123. Ibid.

124. See the article entitled 'The Agricultural Interest', in the *South Australian* (22 January 1847), p.5, cols.3,4.

125. 'Old Colonist' [probably Henry Jones], *Colonists, Copper, and Corn in the Colony of South Australia 1850–51* (No publication date is shown, but 1851 or 1852 appears the most likely).

126. Willunga is approximately 30 miles due south of Adelaide.

127. See n.1 above.

128. At what is now the city of Ballaarat. Later, further rich gold 'strikes' were made at Bendigo and other places.

129. Note, however, that this loss is not reflected strongly in the population figures for the colony (see Table 2), due to a substantial inflow of migrants from England.

130. See E. Dunsdorfs, *The Australian Wheat-Growing Industry 1788–1948* (1956), pp.28 and 175–6.

131. My emphasis (L.J.J.).

132. E. Dunsdorfs, op. cit., p.175.

133. The figures in this Table have been reproduced from E. Dunsdorfs, *The Australian Wheat-Growing Industry 1788–1948* (1956), pp. 532, 534, 475; W. Harcus, *South Australia* (1876), table facing p.394; and (for the period 1876–1884) the appropriate *South Australian Parliamentary Papers*. It should be noted that the export figures in cols.6 and 7 are in each case related to the production figures (cols.3 and 5) for the preceding year. Also, the variations in yield per acre shown in col.4 indicate the type of season experienced — e.g., low values signify drought years.

134. E.J. Crawford to John Ridley, 19 December 1856 (MS., document No. 1053/60 in the Archives of the State Library of South Australia, Adelaide).

135. A. Forster, op. cit. (see n.49 above), pp.463–4.

136. W. Harcus, op. cit. (see n.133 above), p.364.

137. Ibid., p.72.

138. C.H. Bagot to John Ridley (see n.110 above).

139. A (South Australian) patent was finally granted to Marshall in 1878, and his mechanism was made and fitted by several local manufacturers thereafter, with beneficial results.

140. See Frances L. Wheelhouse, *Digging Stick to Rotary Hoe* (1966), chapter 5.

Emile Lamm's Self-propelled Tramcars 1870–72 and the Evolution of the Fireless Locomotive

D.G. TUCKER

I Introduction

Emile Lamm (usually referred to as 'Dr' — he was, in fact, a dentist by profession*) of New Orleans, Louisiana, U.S.A., is seldom mentioned in modern books or articles.[1] He did receive some mention in D. Kinnear Clark's classic work of 1878,[2] but even there no study of his ideas was made. Nevertheless, examination of his thinking over a short period of just over two years, from mid-1870 to mid-1872, is extremely interesting and clearly illustrates an important concept, namely, that a theoretically ideal but complex solution of a problem, giving high efficiency, has, in practice, to give way to a much less efficient but much simpler solution. The matter is examined here through Lamm's patent specifications. Lamm himself published very little, and it has not so far proved possible to obtain even the one paper by him to which reference has been found.[3]

The problem Lamm tackled was how to provide self-propulsion in a street-car (i.e. tramcar) without producing smoke. His work led eventually to the development of the type of fireless steam locomotive so extensively used in industrial applications during the past century, namely one relying for its steam on the heat stored in superheated water under high pressure, the heat being injected at the base station by high-pressure steam fed into the water in the insulated container on the locomotive. Fireless steam propulsion had not, it is believed, been tried in the U.S.A. before 1870, and Lamm was probably quite unaware of Colburn's earlier proposals in Britain,[4] which, however, did not lead to practical trials of a true fireless system.

In the two-year period mentioned, Lamm was granted six U.S. Patents. The systems disclosed in these patents are described fully, in the present author's terms and using simplified schematic diagrams, in

*An interesting short biography is reproduced in the Appendix.

Section II below. There were also two British patents,[5] but these disclosed no ideas not covered in the U.S. patents. The first idea Lamm tried (represented by Figs. 3 and 4) was the use of a conventional steam engine, but driven by ammonia gas instead of steam. The gas was produced by evaporation of liquefied ammonia contained in a vessel immersed in water, and the exhaust gas from the engine was brought back to the water surrounding the ammonia vessel, and dissolved in it, for three reasons:–

1. To recover the expensive gas, which could be easily recovered from the solution in water in the fixed plant at the terminal station, and re-used after compression and liquefaction.

2. To prevent the escape of the noxious ammonia fumes in the tramcar and its surroundings.

3. To recover a large part (Lamm thought all) of the latent heat of evaporation and keep the water and liquefied ammonia from falling in temperature too rapidly — important for practical reasons as well as for efficiency.

The idea was sound. The vertical-tank arrangement of the first patent was evidently unsatisfactory for tramcar use; but the second arrangement (Fig. 4), with horizontal vessels, lent itself to the

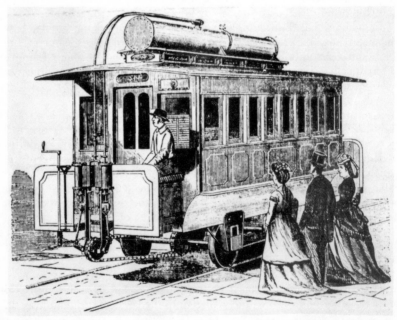

Figure 1. Self-propelled tramcar by E. Lamm, 1871, using ammonia (from *The Engineer*, **33**, 1872).

convenient layout shown in Fig. 1, where the vessels are mounted on the roof, with the engine set vertically at one end of the car. A car, believed to be as shown in Fig. 1, was actually put on trial in New Orleans in 1871 and apparently worked satisfactorily from the mechanical point of view; but the escape of ammonia, in spite of water or oil seals as described in the patents, could not be sufficiently prevented to make it acceptable; and the chemical action of the ammonia on the iron parts was found to be serious.[6] Thus the ammonia-propulsion project was abandoned. Although ammonia engines were not suitable for tramcars, other applications for them were developed later.

It seems that by early 1872 Lamm had realized that steam was perhaps the best medium for converting heat at a fixed station to motive power on a vehicle, and patented the idea shown in Fig. 5. Basically this comprised a pressure-tank of superheated water (refilled with a fresh supply for each trip, as explained in Section II) from which steam was produced as required to drive the engine. The tank was immersed in a larger vessel containing a saturated solution of calcium chloride whose temperature was raised initially to its boiling point, which Lamm states to be at least 380°F (193°C); this keeps the water vessel superheated. The exhaust steam from the engine was passed into the outer tank, where it was absorbed, giving up its latent heat (according to Lamm) and thus keeping the calcium chloride solution hot although it was being diluted by the condensate.[7] No evidence has been found that this system was given a practical trial, and perhaps even its inventor had doubts about it, because shortly afterwards he patented the simple hot-water system of Fig. 6. This comprised just the tank of superheated water, refilled after each trip, the spent steam being exhausted to the atmosphere.

Within a short time, Lamm had discovered a much better way of obtaining the superheated water, namely by passing high-pressure steam into the tank through a perforated tube as shown in Fig. 7. This became a standard system of working fireless steam locomotives,[8] and Lamm received credit for it. The process by which he had moved from the theoretically efficient ammonia system to the successful simple system which wasted the exhaust heat is interesting. It is a fact, however, that the scheme of Fig. 7 had been almost exactly anticipated in Colburn's article of 1864, where he wrote: 'a 4 in. steam pipe, descending vertically into the reservoir, and extending, as a perforated distributor, along the bottom'.

A fireless locomotive made on these principles to Lamm's design for the Ammonia and Thermo-Specific Propelling Company of America is shown in Fig. 2; it was used on the New Orleans–Carrollton tramway in 1872.[9] The starting pressure was 180 lb/in^2; after a 12-mile journey hauling an ordinary tramcar, the pressure had fallen to 50 lb/in^2. (Although Lamm's patent describes the system as for a streetcar, no evidence has been found that he tried it on a self-propelled car.) The

Figure 2. Fireless steam locomotive hauling tramcar, by E. Lamm, 1872
(from *Scientific American*, **27**, 1872).

locomotive was evidently successful, for by 1874 we read that such
locomotives (in the plural) 'are in regular use' in New Orleans.[10]

II Lamm's Patents, 1870–72

1   U.S. PATENT No. 105,581 DATED 19 JULY 1870, 'AMMONIACAL
    GAS ENGINE'

A schematic diagram of the system disclosed by this patent is shown in
Fig. 3. Liquefied ammonia gas is pumped into the vessel indicated,
through the pipe A, until it is full. This vessel comprises a number of
vertical tubes, below the dashed line, so that a large heating surface is
provided. It is immersed in the outer tank which is two-thirds full of
dilute ammonia solution in water; this tank is filled through pipe B and
emptied through pipe C. The top and bottom of the ammonia-gas
vessel are connected to the buffer vessel, from which the gas input to
the engine is taken. The exhaust from the engine, i.e. low-pressure
ammonia gas, is taken by the pipe shown to the bottom of the tank of
ammonia solution, where it dissolves. To prevent leakage of ammonia
gas from the piston-rod glands and from the valve-gear of the engine,

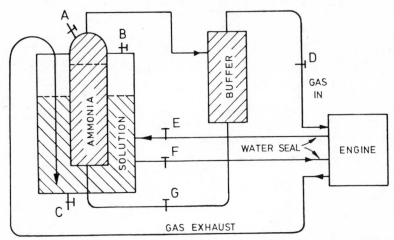

Figure 3. Lamm's ammonia engine of July 1870. (Stopcocks in all diagrams are indicated thus ∓.)

these parts are enclosed in water, which tends to circulate through the outer tank by means of the pipes shown.

The cock D controls the power and speed of the engine. Cocks E and F control or isolate the water seals of the engine. The way cock G is used is not clear.

At the end of the operation, most of the liquefied ammonia has been used and has been transferred into solution in the tank. It is drained out through pipe C and the solution is treated for the recovery of the ammonia as liquefied gas, the tank being refilled with dilute solution. The inner vessel is at the same time recharged with liquefied ammonia gas already prepared. The engine is then ready for the next period of operation.

U.S. Patent No. 121,527 dated 5 December 1871, 'Tender for Gas Boilers', was a small improvement, consisting of the addition of another vessel containing liquefied ammonia gas and connected to the bottom of the main vessel in Fig. 3. The object appeared to be the maintenance of a higher pressure of gas as the operation proceeded.

2   U.S. PATENT No. 124,495, DATED 12 MARCH 1872, 'IMPROVEMENT IN AMMONIACAL GAS ENGINE'

This revised scheme for the ammonia gas engine is shown diagrammatically in Fig. 4. The 'boiler' is now horizontal instead of vertical, and instead of being made of tubes, is an ordinary cylindrical vessel with longitudinal tubes like the water-tubes of an ordinary steam boiler. It is immersed in a horizontal tank of dilute ammonia solution. The buffer vessel of Fig. 3 has been dispensed with. Evidently the

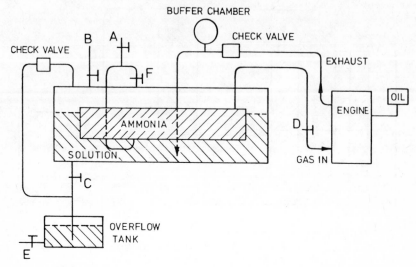

Figure 4. Lamm's improved ammonia engine of March 1872.

water seal of the previous scheme was not found satisfactory, as there is now an oil seal in the engine. The exhaust pipe has a buffer chamber with a check valve in order to relieve back-pressure on the engine valve which otherwise would tend to lift it off its seat at the moment of exhaust. An overflow tank containing ammonia solution is provided, so that if the ammonia gas from the exhaust of the engine does not dissolve quickly enough in the main tank, the excess gas will pass via the check valve to dissolve in the overflow tank. The check valve prevents passage back to the main tank if the pressure in the overflow tank rises too much.

The ammonia tank is filled via cock A, the outer tank via cock B. The ammonia solution is drained, at the end of an operation, by opening cocks C and E, and the process can be accelerated by opening cock F (while A is closed) so that the residual gas pressure in the inner vessel (or boiler) aids the gravity discharge.

3　U.S. PATENT No. 124,594 DATED 12 MARCH 1872, 'IMPROVEMENT IN CHLORIDE-OF-CALCIUM ENGINES'

The equipment on the vehicle was quite simple, as shown in Fig. 5(a). The equipment at the charging station was relatively complex and Lamm evidently felt it necessary to include it in the patent; it is shown schematically in Fig. 5(b).

The inner vessel in Fig. 5(a) supplies steam from the superheated water to the engine; the outer vessel contains a saturated solution of

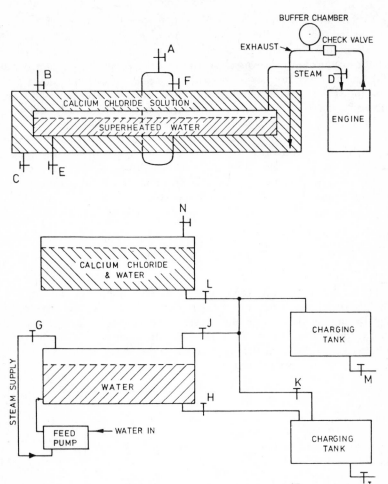

Figure 5. Lamm's 'Chloride of Calcium Engine', March 1872:
    (a) Unit on the tramcar
    (b) Stationary unit for charging mobile unit.

calcium chloride at its boiling point. The exhaust steam from the engine is fed via a check valve and buffer chamber (as in the ammonia-gas engine patent) to the bottom of the calcium chloride solution, where it condenses.

In Fig. 5(b), both the main vessels, one for making the saturated solution and the other the superheated water, are boilers heated by a fire. The water boiler has a feed pump driven by its own steam through cock G. To charge the vessel on the vehicle with superheated water, cock H is opened and the charging tank fills; the quantity is just sufficient for the vehicle. Cock H is then closed, and cocks I and A are connected by a flexible pipe and opened; cocks J and K are then opened, and the steam pressure drives the superheated water from the charging tank into the vessel on the vehicle. A similar process transfers the calcium chloride solution to its charging tank using cock L; cocks M and B are connected by a flexible pipe and opened, and steam pressure via cock J transfers the solution to the outer vessel on the vehicle.

At the end of the operation, cocks C and N are connected and opened, and steam pressure via cock F (A closed) drives the weakened solution back to the boiler for concentrating again. Although not mentioned in the patent, cock E is presumably used for draining water from the vessel at the end of each operation.

4   U.S. PATENT No. 125,577, DATED 9 APRIL 1872,
    'IMPROVEMENT IN SUPPLYING STEAM TO
    TRAVELING ENGINES'

This is shown schematically in Fig. 6. The system is very simple. A single vessel contains superheated water which supplies steam to a steam engine. To charge the vessel when empty, it is first heated by connection, through cock A, to the steam from the boiler at the charging station; then it is connected to the water in the boiler so that it fills to the required depth.

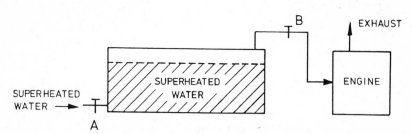

Figure 6. Lamm's fireless steam engine, April 1872.

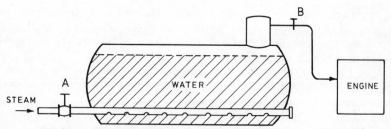

Figure 7. Lamm's improved fireless steam engine, July 1872.

5 U.S. PATENT No. 129,969, DATED 30 JULY 1872,
   'IMPROVEMENT IN SUPPLYING STEAM TO
   TRAVELING ENGINES'

This is shown in Fig. 7. It differs from the scheme of Fig. 6 in one important way: instead of the vessel being recharged with superheated water at the beginning of each operation, the water is not removed from the vessel, but instead is re-superheated by passing steam into it through cock A and the perforated tube in the bottom of the vessel. The total area of the perforations should equal the cross-sectional area of the tube.

This patent was re-issued with expanded description and claims as Re-issue No. 5083, dated 1 October 1872.

III   The Development of Fireless Locomotion during the 1870s

Lamm seems to have made the first-ever fireless steam locomotive in 1872, as described in Section I. Other people, however, were quick to copy it. How Lamm's patent protection was used, if at all, is not clear.

Trials of two fireless locomotives took place on the tramway between East New York and Canarsie (3.5 miles) in October 1873[11] and again a month or two later.[12] They were made by the Fireless Engine Co. (president G.L. Laughland, consulting engineer C.H. Haswell). One locomotive did the return trip in just under 30 minutes' running time, hauling a tramcar carrying 120 passengers, with steam pressure falling from 180 to 45 lb/in². Its weight was 4 ton 3 cwt; it had condensers and worked without expansion. The second locomotive was tested by R.H. Buel and H.L. Brevoort; it seemed to be less successful, taking 35.5 minutes to cover 4.4 miles, with steam pressure falling from 142 to 22 lb/in². This locomotive certainly used the system of Fig. 7, and the reports inferred that both locomotives were constructed to Lamm's design.

In 1875 a fireless steam tramcar was made by Bède and Co. in Belgium for use by the Société Générale de Tramways.[13] The hot-water vessels comprised four horizontal ones, placed under the seats, and two upright ones, totalling 50 cu. ft., all well-lagged. Three cylinders, 4.5 in. by 14.2 in., drove on to one axle. Initial steam pressure was 162 lb/in², and the car could run for about 50 minutes. It was apparently kept in use for a year or two, but was found expensive to operate.

Also in 1875, details of a fireless steam tramcar designed by L.J. Todd, of Leith in Scotland, were published.[14] A drawing of the car is shown in Fig. 8, and an exterior view of it, showing the arrangement for charging the hot-water reservoir, is reproduced in Fig. 9. Clark[15] states that the car was constructed, but there is room for doubt on this point, because, while the article in *The Engineer* is so worded as to *infer*

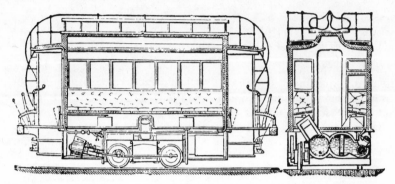

Figure 8. Todd's fireless steam tramcar, 1875 (from D.K. Clark, *Tramways*).

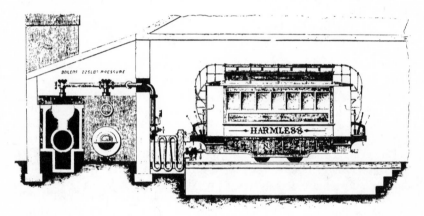

Figure 9. Todd's fireless steam tramcar, 1875, showing charging arrangements
(from *The Engineer*, **39**, 1875).

that the car was actually in existence, it carefully avoids definitely
saying so; and furthermore, no reports of its trials have been found.
Todd was, however, an experienced designer of tramway vehicles and
locomotives which had actually been constructed, and he had
approached the idea of the fireless steam engine quite logically, but by
a process of thinking very different from Lamm's. He set out his ideas
very fully in 1874,[16] and made it clear that he had already tried out a
tramway locomotive which was designed to run mainly on stored heat
in an extra-large boiler, the very deep fire being used to generate this
heat mainly while the machine was waiting at the terminus, and closed
up during the journey, so that no smoke would be produced. The
progression from this idea to the fireless steam machine was natural,
although he generously credited it to Lamm.

Todd made a patent application in 1875 which clearly describes the basis of the fireless tramcar:[17]

> Relates to tramcars, omnibuses, etc., which are propelled by steam generated in stationary boilers and stored in receivers on the carriages. In one arrangement, two receivers, placed longitudinally beneath the floor, have at one end a cylinder and piston arranged to drive two coupled pairs of wheels, and at the other end an extension to balance the weight of the cylinder, domes or steam spaces being fitted to the receivers so as to project up laterally under the seats. The cylinders have jackets supplied with water or steam from the receivers.

It seemed a sound and practical design.

Of all the work done on fireless locomotion during the 1870s, the best documented is that by Léon Francq of Paris. His system was essentially that patented by Lamm in 1872 (Fig. 7). It is curious that

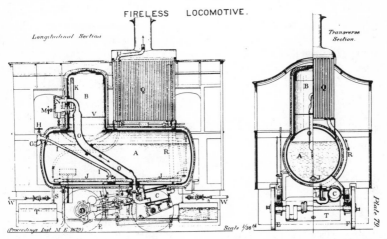

Figure 10. Francq's fireless tram locomotive built by Hunslet Engine Co. in 1879 (from *Proc. Inst. Mech. Eng.*, 1879).

A   hot-water reservoir, holding 400 gallons, 220 lb/in^2 pressure
B   dome
C   cylinders
G   nozzle for introducing high-pressure steam
J   perforated pipe for steam entry to water
K   steam outlet
M   valve
N   expander, to deliver steam to cylinders at constant pressure
O   steam-pipe to cylinders, passing through reservoir to provide heat to dry the steam
Q   air-cooled condenser
S   pipe to feed condensate to tanks T
U   exhaust tube for any steam not condensed.

in an account of Francq's work in 1879,[18] it is stated that Francq 'introduced the improved plan of reheating the water in the reservoir by the injection of steam at high pressure', whereas in Lamm's scheme 'the reservoir applied by him to his locomotive was recharged with water (heated under pressure) at suitable intervals'. Why this myth arose is not clear, since Lamm's locomotive of 1872, according to Clark's description already cited in Section I, quite clearly had its water heated by steam-injection. Nevertheless, Francq obtained patents.[19] His earliest fireless locomotive, of 6.5 tons empty, was tried on the tramway between Saint-Augustin and the Boulevard Bineau,[20] and did the round trip of 5 miles with pressure falling from 156 to 50 lb/in^2. The exhaust steam proved an annoyance. Later locomotives had condensers as well as the higher pressure of 213 lb/in^2; these performed well on the tramway between Rueil and Marly-le-Roi (near Paris), where over 10,000 km were run each month. A rather similar tramway locomotive was built for Francq by the Hunslet Engine Co. of Leeds in 1879, but it is not known if it was ever put to work.[21] A full account of his work was given by Francq in 1879 to the Institution of Mechanical Engineers,[22] and from this the drawing of the Hunslet locomotive is reproduced as Fig. 10. Some account of Francq's later work on fireless locomotives is given by Clark.[23]

By this time the fireless locomotive had been developed to substantially its final form.

IV Conclusion

It can be seen that the decade of the 1870s was the decisive period for the evolution of the fireless steam locomotive, and that, even if Emile Lamm was not the first to have the basic idea, he certainly showed great originality in his approach via the ammonia engine and was almost certainly the first to demonstrate a true fireless self-propelled vehicle or locomotive. His abandonment of theoretically efficient but complex systems for a less efficient but simple system probably illustrates a principle of general validity. It is perhaps a matter of some surprise that a dentist of that period, with, as far as is known, no scientific or engineering training, should not only be able to invent and design effective and complex machines but should also be able to apply to the process of invention an understanding of quite subtle scientific matters.

V Acknowledgements

I am very grateful to Mr J.A. Crabtree and Mr J.G.B. Hills for discussion and comment on this work and for valuable information and references; and also to several libraries, notably those of the Science Museum and of the Institution of Civil Engineers.

Appendix

BRIEF BIOGRAPHY OF EMILE LAMM FROM *APPLETON'S CYCLOPAEDIA OF AMERICAN BIOGRAPHY* (NEW YORK, 1887)

LAMM, Emile, inventor, b. in Ay, France, 24 Nov., 1834; d. near Mandeville, La., 12 July, 1873. He was educated at the Collège royal in Metz, but came to the United States in 1848, and became a dentist, following his profession in Alexandria, La., until the civil war. Dr. Lamm served in the Confederate army under Gen. Braxton Bragg during the war, and after its close resumed his practice in New Orleans. As a boy he showed decided mechanical ingenuity, and in 1869 devised an ammoniacal fireless engine for the propulsion of street-cars. The system was tested by street railway companies in New Orleans, New York, St. Louis, and other cities, with satisfactory results; but owing to Mr. Lamm's premature death and unfortunate management on the part of the company that controlled the patent, the motor has not been put into practical operation in the United States. The system has been introduced in France and Germany, where it has been improved and perfected, so that at present (1887) it is extensively used for street-cars and vehicles. During his work on this invention he became impressed with the facility with which the vapour of water may be condensed, even at an elevated temperature, in water under high pressure; and pursuing his experiments, he produced another fireless engine, which he patented in 1872, and which is now in practical use. He also invented a method for the manufacture of sponge gold, for which he obtained a patent and a gold medal at the Mechanics' fair in New Orleans. This process is used largely by dentists throughout the United States. Dr. Lamm was a fellow of the New Orleans academy of sciences. He was drowned.

NOTE by the present author: The statement above that the ammonia engine had been successful and used extensively in Europe is contrary to D.K. Clark's testimony and is not supported by any references I have found.

Notes

1. An exception is R.J. Buckley, *A History of Tramways* (Newton Abbot, 1975), where Lamm gets several lines. One might expect more in A. Baker and A. Civil, *Fireless Locomotives* (Tarrant Hinton, 1976), but in fact there is less.

2. D. Kinnear Clark, *Tramways, their Construction and Working* (London, Vol.1, 1878; Vol.2, 1882; 2nd edition, 1894).

3. E. Lamm, 'Locomotive without fire', *American Artisan*, 3rd series, 3, c.1871 or 1872, p.312. This reference is given in Bruno Kerl, *Repertorium der Technischen Literatur, 1869–1873* (Leipzig, 1878), p.790. English-language bibliographies of periodicals so far consulted do not list the *American Artisan* before 1880.

4. Z. Colburn, 'On the working of underground railways', *The Engineer*, 18 (1864), 321–3 and 335–6. Colburn was not concerned with tramways, but rather with the Metropolitan Railway in London.

5. A. Armour (communicated by E. Lamm), British Patent 517, dated 25 Feb. 1871; and W.R. Lake (communicated by E. Lamm), Brit. Pat. 903, dated 25 March 1872. A curious point is that the ideas disclosed in the second British Patent *antedate* the corresponding U.S. Patents Nos. 125,577 and 129,969 discussed in Section II of the paper, in the second case by over four months.

6. Clark, op. cit., 1878 edition, pp.316–17; 1894 edition, pp.413–14; also see 'Ammonia as a motive power for street cars', *The Engineer*, 33 (1872), 23–24. On the question of corrosion of the iron parts by ammonia, Mr J.G.B. Hills informs me that: 'Practical ammonia engines, with iron parts, have been made since Lamm's time. They were formerly used in ice works to recover part of the energy used in the gas compressors. Ammonia vapour, and the liquefied gas, do not corrode ferrous materials (they are often handled in mild steel pipes and vessels in chemical and refrigerating plants). In the presence of water or water vapour, they are severely corrosive. Copper alloys are corroded wet or dry, and have to be avoided.' The trials in New Orleans were described in *Scientific American*, 25 (1871), 290–91; this article mentioned that the Ammonia Propelling Company had been formed to exploit the invention. Reference was also made to a quite different kind of ammonia engine invented in France in 1865 by Ch. Tellier using the vacuum produced in a cylinder filled with ammonia gas when water is injected. The reference given was to *Trans. American Institute*, 1865–6, p.436, but I have been unable to trace this journal for the years quoted. Two brief relevant notes were published in *Comptes Rendus Acad. des Sciences*, Paris, 60 (1865), 59 and 1195.

7. Although at first sight it seems unlikely that steam could give up its latent heat to a liquid hotter than itself, the principle is supported by O. Lyle, *The Efficient Use of Steam* (H.M.S.O., 1947), p.790, and was apparently used in practical locomotives by Honigman at Aachen about 1885.

8. Lyle, cited in the previous reference, discusses methods of blowing steam into a liquid at pp.354–6; unsuitable methods caused much noise, vibration, and disintegration of plant. He says: 'Perforated pipes are often used, sometimes quite successfully. They often fail, however, and sometimes hammer badly. . . . If a perforated pipe is to be used it should be long with small holes well spaced.'

9. 'Fireless locomotive', *Scientific American*, 27 (1872) p.118; Clark, op. cit., 1878 edition, pp.317–18; 1894 edition, pp.414–15.

10. 'The fireless locomotive', *The Engineer*, 37 (20 Feb. 1874), 135.

11. *The Engineer*, 36 (24 Oct. 1873), 276.

12. *The Engineer*, 37 (20 Feb. 1874), 135.

13. Clark, op. cit., 1878 edition, pp.336–7; 1894 edition, pp.430–1. The relevant patents are Belgian Patents Nos. 36129 and 38008 of 1875.

14. 'Todd's self-propelling tramway car', *The Engineer*, 39 (9 April 1875), 240 and 243.

15. Clark, op. cit., 1878 edition, pp.334–5; 1894 edition, pp.412–13.

16. L.J. Todd, 'On working street railways by steam power', *The Engineer*, 38 (24 July 1874), pp.65–71.

17. British Pat. No. 355, dated 30 January 1875. This patent was filed only as a provisional application; the quotation is from the official abridgement.

18. 'Fireless locomotives', *Engineering*, 28 (17 Oct. 1879), 306, plus 2pp. of plates.

19. e.g. French Patent No. 108,954, dated 22 July 1875; Belgian Patent No. 37566 dated 2 August 1875; Brit. Pat. No. 2253, dated 6 June 1878. It is a matter of interest that in published accounts of Francq's work on tramways a decade later, his fireless locomotives were described as being on the 'Système Francq et Lamm'. Thus Lamm's contribution was still being recognized by Francq fifteen years after Lamm's death. See (i) 'Traction au moyen des Locomotives à Vapeur sans Feu', reprint from 'Rapport de la Société Indo-Néerlandaise de Tramways', (Paris, 1888); and (ii) 'Traction et Moteurs à Vapeur sans Feu', booklet published by La Compagnie d'exploitation des Locomotives sans Foyer (Paris, n.d.); both in the Library of the Institution of Civil Engineers.

20. Clark, op. cit., 1878 edition, pp.332–4; 1894 edition, pp.509–10.

21. A. Baker and A. Civil, *Fireless Locomotives* (Tarrant Hinton, 1976), pp.52–3. Plate 34 of this work shows a photograph of the Hunslet locomotive.

22. L. Francq, 'On fireless locomotives for tramways', *Proc. Inst. Mech. Eng.* (1879), 610–26, plus plates 79 and 80. The discussion on pp.626–41 is also interesting.

23. Clark, op. cit., 1894 edition, pp.512–13.

British Industry in 1767: Extracts from a Travel Journal of Joseph Banks

Transcribed and edited by
S. R. BROADBRIDGE

Both the text and the editorial matter that follow have been abbreviated from a complete edition of Banks's 1767–68 travel diary prepared by the late Mr Broadbridge, which still awaits publication

Introduction

The journal from which extracts are printed here for the first time was written by Joseph Banks (1743–1820) when he was twenty-four years old. Since his father died while Banks was at Oxford (1760–63) he entered on his patrimony of at least £6,000 a year when he came of age in 1764; from then onwards there was no financial obstacle to the pursuit of his interests which while centred on botany, widened to include most aspects of science and technology.[1]

In his early adult years he was seldom in London. In April 1766 he accompanied Sir Thomas Adams in the *Niger* as naturalist to his expedition to Labrador and Newfoundland.[2] Soon after he set out he was elected a Fellow of the Royal Society; he was twenty-three.

He returned at the end of the year and spent two years in England. In May and June 1767 he toured the Bristol area keeping a journal which, after considerable vicissitudes, was published in 1899.[3] Compared with the present journal it adds very little to our knowledge of the area or the times. In August he set out on the journey which is partly described here. At the commencement his notes were mainly botanical but they soon developed much more widely.

Later in 1768 Banks planned a visit to Lapland, but instead joined the expedition to the South Seas led by James Cook, in order to observe the 1769 transit of Venus. Subsequently, finding himself unable to join Cook on his second voyage, Banks made a last voyage to Iceland in 1772. He became, as President of the Royal Society from 1778, the dominant figure in English science. He was made a baronet in 1781.

Banks's manuscripts have been widely scattered and even the famous *Journal* of the South Seas voyage has only recently been

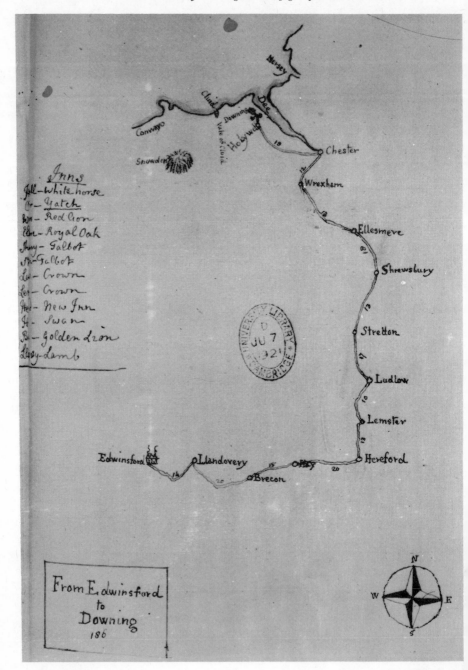

Inns
Jll — White horse
G — Yatch
Wm — Red Lion
Eln — Royal Oak
Shny — Talbot
Sh — Talbot
Lu — Crown
Les — Crown
Her — New Inn
Sf — Swan
Br — Golden Lion
Lway — Lamb

From Edwinsford
to
Downing
186

properly published. The journal from which extracts of technical interest are printed here is in the Cambridge University Library,[4] entitled *Journal of an Excursion to Wales &c Begun August ye 13th 1767 Ended January ye 29th 1768.* It is written in Banks's own hand on quarto paper. The entries were clearly written up at intervals rather than day by day.

Travelling via Oxford and Gloucester, Banks spent many weeks at Edwinsford, Carmarthenshire, with friends, during which time he made side-trips to such places as the Roman gold-workings at Gogofau and also dug a barrow. On 16 November he set out through the border country to Chester; from the 21st to 3 December he stayed with Thomas Pennant at Downing (see map), visiting lead-mines, white lead works, and the herring fishery. He spent early December on his own estate at Cheadle, Staffs., and then moved north to see the Duke of Bridgewater's Canal, taking in the Potteries on the way. He visited coal mines, lime quarries, and salt mines; he saw Kay's flying shuttle, which had begun to come into use about ten years before. Then he moved south once more, to Lillieshall, Salop, and Coalbrookdale which he described in some detail, and after a look at Wednesbury (iron) and Coventry (ribbons) he returned home to London.

The Journal

[28 August 1767] Harvest is now beginning a great deal of corn of all Kinds in this Countrey [Carmarthenshire] but the weather for getting it in but indifferent the Bromus secalinus [Rye-brome] is very common all over this countrey among their wheat it is calld in welsh Llar the cradle Scythes[5] for the introduction of which a premium was given Lately by the Society [of Arts] by the name of Hannault [Hainault] Scythes have been used here time immemorial they mow all their oats & barley with them & lay it smoother & evener for binding than it can possibly be done by a sickle all their corn is Bound in very small Sheaves & immediately piled into what they call field mows a method of husbandry so much superior to all others that I cannot but wonder why it is not universal . . . they say in this Countrey that these mows will bear a month or Six weeks Rainy weather without the Least damage their method of Raking in this Countrey is I think much superior to ours the heads of their Rakes are 6 feet Long with strong & long teeth the handles of about the same length have a pin stuck into them about 2 feet from the End which the right hand Lays hold of while the Left guides it two or three or maybe more people go in a row with these rakes trailing after them discharging what they pick up in Equal rows which if they Keep together they cannot fail of Doing by this means the most part of the Corn is Collected and what the Rake misses is Laid straight so that a crossraking is sure to pick it up in this manner if Land is carefully managed tho a great deal of corn is left in the Binding it will be so compleatly got together that scarce enough will be left for the pigs & geese & that too in half the time & with half the trouble of using our light rakes

[25 September 1767] From hence [Llandilo] nothing remarkable occurs till you come to Aberguilly [Abergwili] the Road indeed every where overlooks a fine Countrey Abeguilly itself is the seat of the Bishop of St. Davids & is the Cheif place where the Salmon fishing of Towey is carried on the Boats used in this fishing are perhaps the most ancient & Least artificial of any used in this Island perhaps in the world as all the canoes I have seen Even of the most unpolished Indians are made with infinitely more art than these they are but corracles their shape is almost round at Least very tubish for a Boat that which I measured was 4 feet 6 by 3 feet 3 their Edges are watled together With small Rods such as baskets are made of in these are interlaced spit sticks such as hoops are made of seven or eight of which Cross the Bottom over these is put a peice of flannel Dipd in tar which is sewd to the watlings the bench is of Deal pretty strong & goes across the middle of the Boat these which are just capable of holding a man are guided by a small paddle consisting of a blade & handle just Like the Indian ones with this the man moves the Boat by putting it in before him & moving it to him two of them will hawl a Salmon net of a very Large size & draw any Pool in Towey

Carmarthen the Capital of this Countey is situated on the Towey which is navigable up to the town for Large vessels its Castle has been strong & large but is now entirely in Ruins. . . .

[Brecon, 16 November 1767] I have here a little time & a little paper which I shall fill up with some account of the Carriage used here calld a car which I believe is the oldest & simplest construction of a carriage that is in use anywhere

it is Composed of two shafts in Lengh about 11 feet upon which are naild 5 of 6 cross barrs in Lengh $3\frac{1}{2}$ which serve to support the weight which is to be carried which is hindered from slipping off behind either by three half loops interlaced & fastened into holes in the Last bar or by two prongs of wood fastening in the shafts with Cross peices between them the ends of the shafts are Sloped off & Trailing upon the ground serve instead of wheels the harness consists of a saddle with a nitch across the middle into which a wythe twisted & fastend into the fore part of the shafts hitches the horse has also a collar of Straw & another of wood over it into which are fastened two rings of Iron Large Enough to receive the End of the shaft which being put through the ring is hindered from returning by a small peg of wood which fastens into a hole made for that purpose

[3 December, 1767] this morn set out for Staffordshire went to holiwell [Flintshire] by the Sea side as beautiful a ride as can be seen in the road had an opportunity of seing a White Lead manufacture which is

caried on thus Pots are provided made of Earthen ware unglazed which
are larger at top than at the Bottom of this shape (Figure 1) Bottoms of

Figure 1.

these are filld with very Strong vinegar[6] then the Lead being Cast into
Plates about the thickness of a shilling is rold into rolls of size & shape
to fitt the upper or Larger part of the pot but care is taken not to roll
them too Close & they are put in with their Edges downwards that the
steam of the vinegar may have free access to all their sides These potts
so filld are pil'd with hot horse litter between Each Layer in a large
square room which is calld a stack here they are sufferd to remain till
the different accidents of heat & c: make the operator Conclude that
they are Enough corroded when they are taken out & the plates of
Lead dryed Some of which are now totaly Converted into White Lead
others have in them a thin plate of Lead still metallick but scarce
worth notice these are pass'd between two Iron rollers whose teeth
meet within one another (Figure 2) which means the white Lead is

Figure 2.

separated from the Metallick part & falls through a Skreen placed
under these rollers while the Lead & any accidental filth is seperated
from it

The white Lead is now Passed through two pair of Stones the
Second set finer than the first & sometimes it is again put through the
fine Stones by which it is made fine enough for use it is then washd by
a very ingenious Contrivance a number of square troughs which they
Call Backs are set in a room each having a Communication with the
next to it at one end of these is a Large round tub in which is a Large
pole which is turnd by water & furnishd at the Bottom with Cross
barrs like a chocolate mill at the other End of the Backs is a pump
workd likewise by water this raises water into the round tub which by
that means being filld too full a quantity runs over again into the Backs
& with it the white Lead Kept in Motion by the wooden pole for that
purpose & thus it is circulated about till the Lead is Enough washd
when it is allowd to Settle the water is then drawn from it & it is taken

up with an Iron Ladle & Laid down upon stone Shelves in Cakes where is soon becomes dry & fit for use (Figure 3)

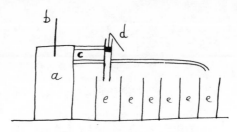

Figure 3.

a   the round vessel where the Lead is washd
b   the pole turnd by water & made at the bottom like a chocolate mill
c   the spout by which the water & Lead run over into the Backs
d   the pump by which it is returnd
e e e back or square tubs communicating with one another

[5 December 1767] From hence [Tarporley, Cheshire] to Namptwich [Nantwich] 10 the Countrey & roads much Like the Last stage Lay here tonight & in the morn went to see the Salt works for which this place is famous it is made intirely from the spring as they have no salt rock here which they complain much of as the Places where the Rock is can boil so much more salt with Less fire that they undersell these much[7] their Salt is however Exceedingly white & good they make of two Kinds Coarse graind & what we call Basket the preparations of which are nearly the same as it consists only in Evaporating the Brine in a Leaden Trough till the Chrystals form & Precipitate themselves when they are taken out & put into Baskets to Drain the only difference in the two is that to make the fine salt the water is made to Boil as feircely as possible all the time but in the Coarse is Gently Evaporated They have but one brine pit open (tho they say that there are springs for some miles up & down the river) which they Call old Byat & have a tradition that it was built by Julius Caesar it is Coverd with a shed of wood Exceedingly shabby & full of holes the salt here is Sold from ninepence to Elevenpence a bushel exclusive of Duty which is three shillings

ECTON [Staffs.]

[9 December 1767] Went to see a Coppermine belonging to the duke of Devonshire[8] probably the richest in the Island this year it will Clear about 11 thousand pounds besides doing the whole years work at a navigable drift which is workd in dead stone without a spark of oar Every circumstance of this mine & the hill it is in are so wonderful that they merit a very particular description the hill itself is situate upon the Banks of the river manifold & shows Where ever the water or any other Cause has bar'd the strata of the Earth a Confusion scarce to be described in one place by the river side the strata make an appearance not unlike this (Figure 4) the first is an arch of about forty feet

Figure 4.

diameter as it goes Lower the bedds get more & more pointed till they are quite sharp by the side of these stand two or three bedds of limestone set right up on their heads so that till they are examined they appear like a vein beyond these the beds are bent but in an inverted arch so that more variety of Positions in the space of 60 or 70 feet can scarcely be conceivd but irregular as this may appear to be the inside of the mine is still more so what they are working upon seems to be a Pipe which instead of going horizontal is set upon its head & goes directly perpendicular it is of an immense size & very irregular figure so that the chambers of the mine are so magnificent as scarcely to be Equald by the finest buildings. The Contents of this pipe are full as irregular as the rest of the hill Limestone Chirt Spar Copper (tho much the largest Quantity of the Latter) Jumbled together in a Confusd mass now a lump of Copper half a ton weight sometimes a lump of Limestone as big but the peices are generally of a size much inferior to this nor do these lay at all regular but turning winding & Twisting all manner of ways as if Barrows full of these materials had been thrown irregularly into a Hole made to receive them

The depth of the mine is 160 fathom from the Crown of the hill from which place it has been brought down by shafts began about King Williams time but never very rich till within these ten years at the depth of 96 fathoms I found plenty of fungi of two different Kinds neither of which I had Ever observed before . . . in one of the drifts also at a great depth issued a small quantity of Bitumen tho but a drop the Miners seemed / Quite unacquainted with it asking me whether I had Ever seen such a thing before & what it was

Here is also found a Small Quantity of a whitish ore much resembling if not realy antimony but this in such small Quantities that a little bit now & then which is Kept as a Curiosity is all that they can get

[12 December 1767] have continued my walk round the Parish doing things not worth remembering here . . .

[18 December 1767] Set out for Worsely [Worsley, Lancs.] with Mr. Gilbert[9] that I might see the Duke of Bridgewaters navigation The road, through Burslem famous for Pottery indeed in this Countrey for seven or Eight miles together Every village is full of Potteries making different Kinds of what is Calld Staffordshire ware

The Cheif materials of these wares especially of what is calld stone ware[10] & the yellow ware[11] now so much in fashion are flints & a Kind of Clay which is brought from devonshire as the flints are from Kent yet both these materials are brought here from so great distances on account of a clay found here of which they make what they call sagots[12] a Kind of bowls in which their wares are put when they go into the furnace that they may be screend from the heat of the fire This Clay is always the warrant of bottom of a coal is when raisd like stone of a colour from light to darkish blue & full of the impressions of Plants it soon melts in the air & when put into the Furnace stands the fire excellently.

The manufacture is very ingenious some parts of it I shall mention Plates & dishes are made on a mould of Alabaster burnt or what we call Plaster on this the patterns are formed but they are afterwards before they are burnt Polish'd & compleated by Lathes & sharp tools which the Clay is very soon firm enough to bear — Handles of Cupps Teapots &c: are made by a box in the Bottom of which is a hole properly shapd to answer the pattern intended the top of this is moveable & by the Power of a strong screw presses the Clay that is in the Box through the Hole which consequently takes the impressions its shape & size has given to it

The Flints used in this manufacture are after being calcind made into Paste in mills[13] for that purpose they are composed of a Large Trough pavd at the bottom with the hardest stone that Can be got in the Center of this is an upright turnd round by a water wheel from the sides near the bottom of this go four horizontal poles each of which drives before it one or more Large stones the flints to be ground are put between this & the pavement & coverd with water

Mr. Wedgwood has lately introduc'd into the manufactory the use of Engine Lathes which work upon the Clay with the greatest Ease not requiring the tool to be fastned to the rest by a screw as is common but held fast by a finger of the workman on the top of the rest roughened like a file

From here went to the place where the famous Trunk or Staffordshire navigation[14] is to be opend through an immense hill Calld Hare Castle a mile & $\frac{3}{4}$ in Length the drift is carried about 100 yards is well archd & nobly sizd but their mortar is so soft & seems to have so little care taken in making it that I cannot help having my fears of accidents which may befall it when it comes to bear a large weight of hill

Lay this night at Knotsford [Knutsford] in Cheshire a small town remarkable for nothing in the morning quickly reachd the dukes navigation [Bridgewater Canal][15] now got to a place calld Dunham[16] the Seat of Lady Stamford where the workmen are now Carrying it upon a Bank over a deep valley which is done at a Large Expense by making a cut in the valley & raising its bottom and sides by Large quantities of Earth so that that the whole navigation with the boats upon it rises gradualy to the point desired the use of which is very manifest as the soil for doing it is Carried by the boats at a very Light Expence which by Land Carrage would have required an immense one they are of two sorts (I mean those used for raising the Bottom) & of an ingenuity of Construction which Everyone must [admire] Those calld Fly boats which are most used are constructed by Joining two Boats together at the distance of about three feet which is done by Cross beams of timber upon these is fix'd a trough whose bottom about two feet ½ wide is fixd immediately over the open Left between the two boats but gradualy increases to the top which is twice that size or upwards (Figure 5) the Bottoms of these troughs are Compos'd of a

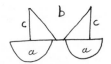

Figure 5.

aa boats   b trough   cc supporters

number of small doors each of which is supported by a chain fastned to it on the inside which passing over the upper side fastens to the side of the boat in such a manner that it is Easily Let to Slip (Figure 6) these

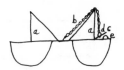

Figure 6.

aa beams by which the sides are supported   b chain
c pin   d ring   e loop

are Loaded with earth from stages & when drawn to any place in the canal to be filld up are with great ease discharged of their Burthen by Letting slip the chains which fasten up the doors & of Course Let every thing out of the bottom in an instant

The others which are calld here diving boats are upon the same principles only they are single boats divided into Partitions one half of which have doors in their Bottoms while the other half are solid & render the boat Boyant after having discharged her cargo

From hence rode to Worseley [Worsley, Lancs.] by the side of the navigation particulars of which I shall mention by & by. . . .

This morn went to see the Mortar mill by which the Mortar usd in the Buildings belonging to the navigation is ground at two operations the first of which is the Breaking it fine after it is Slackd & mixing it with water this is done by a millstone turnd in a Horizontal position in the middle of which is placd a Hollow stone about ten inches in height (Figure 7) on the surface of this two Large Grinding stones set upon

Figure 7.

a millstone   b small round stone

their Edges are turnd following Each other round the stone in the middle under these the Lime is put (which has been carefully Kept in an archd room from the time that it was slackd till now) & ground dry till all the small Lumps are reduced to a fine powder & any any Large ones which may have accidentaly got in are taken out water is then put to it & mixd till it is wet enough when it is removd into the Blending mill there to be mix'd with Sand This is made by a Cylinder of Cast Iron fixd Horizontaly & turned by water it is open at one End & in it are fixd four Iron Barrs running paralell with its sides about one third distant from them and two from the Center (Figure 8) in this are put

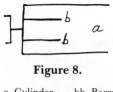

Figure 8.

a Cylinder   bb Barrs

mortar & sand in different proportions according to the uses they are intended for which while it turns round drop from side to side & being cut & torn in its Passage by the barrs are blended & mixed in a more perfect manner than could otherwise be done

The sand of which this Mortar is made is also Carefully washd &

riddled so that no lump nor any Kind of mould can remain in it & thus prepard certainly is stronger & setts quicker either under water than any other in the Island

[21 December 1767] Set out this Morn for Wigan in the road went into a Coal pit where I might with safety view the fire-damp Issue from the Earth it rises like a spring but not with quite so constant a motion bubling violently then resting a second or two then bubling again on a Candle being applyd to the bubbles as they rose out of water they took fire as they burst much like the vapour of spirits of wine burning for an instant & then going out

Lay tonight at Wigan an old town famous now for nothing but the inhabitants say that in the beginning of the Last Century almost all the Pewtery & Braziering in the Island was here

[25 December 1767] this Evening Late [Christmas Day] set out for Rochdale in the way to some allum works which we mean to See tomorrow got there at 8 o'clock

[26 December 1767] got up this Morn Early to see the Famous Shuttle[17] with which Broad cloth is woven by one man sitting still which was used to be done by two or more throwing the shuttle from one to another its contrivance is simple & ingenious (Figure 9) across

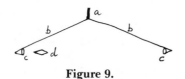

Figure 9.

a handle   bb Lines   cc sockets   d shuttle

under the web is a groove on which the Lower most threads rest upon this the shuttle runs two sides of which are set upon friction wheels the man who setts in the middle of his webb moves it by strings which Communicate with each end of the groove and have at their Ends a socket which fitts well the point of the shuttle in the middle is a handle of wood by Jerking of which the man gives motion to the shuttle which as soon as it is thrown from one side catches in the socket on the other & is by that returnd with another Jerk & so backwards & forwards with more ease than a man can weave a standard web

This history of this shuttle is remarkable a poor weaver invented it about twenty years ago & to secure to himself the profits of so useful an invention procurd the Kings Letters patent for the sole use of it [in 1733] he then made it & all the manufacturers were glad to use it paying him a consideration this Lasted for some time till tird of Paying they all agreed to make a publick purse to defend themselves and to refuse to pay him any thing this was accordingly done the inventor sued them & spent what Little money he had without being able to gain any releif he then tird as he well might be with ingratitude resolvd to set out for France with his invention this he accordingly did receivd a large premium from the French government for it & now Lives among them no doubt teaching them Every branch of our much valued manufactory[18]

The excellence of the invention I dwell more upon as the gentleman who shewd it to me told me he himself had Known as much given for weaving a piece of Cloth only as he now gives for both spinning & weaving

From this place went through very bad roads to blakebourn [Blackburn] about three miles from whence the allum workes are situate[19] to these we immediately walkd on foot The Rock of Allum stone is of great thickness which is nescessary as vast quantities of it are used to make a small quantity of the Salt it is Black & resembling not a little Very Coarse Earthy shale but its cheif Characteristick is its taste which is very strong of allum especialy in those parts which are exposed to the air it is all got by open work & Lay not long ago immediately under the vegetable soil but they have lately met a small Fault which has thrown a cap of Gretstone about 8 yards upon it which they complain much of & say that it takes away much of their profit in the expence of removing it

For making the salt a large heap of the Allum rock or Shale is pild up with a few faggots of furze or brush wood put under these being set on fire communicate to the stone which has so much of the Coal quality in it as just to burn & by very slow degrees calcine itself through

when sufficiently calcind it is ready to yeild its salts & is therefore removd into troughs of water for that purpose where it is steepd till the water has imbibd salts Enough to be ready for boiling it is then pump'd up into Large Leaden boilers & a very quick fire being put under it is Evaporated till ready to remove into the strikeing tubs which are Let into the ground & made of wood like tun vats here it is mixd with Kelp lye & urine & attended by the striker whose business is a profound secret by him it is caused to shoot into small Chrystals which are taken out washd & again dissolved in a small proportion of water & pourd into large casks to the sides of which it fixes itself in Large Chrystals & is then Saleable Allum bringing to the proprietory at the time about 20 pounds a ton

The inside of a Cask when the head is first struck out is the most beautiful sight imaginable resembling a Cavern filld with Large Chrystals pointing all ways at this time more than semi transparent & in all the different positions that can be conceivd

[27 December 1767] Returnd this day to Worsely through as Cold weather as I ever felt roads not the best

[29 December 1767] went by the Level[20] into the Dukes Coals in about

7 hours with some fatigue saw the Greatest part of them They are got by what is calld stall or bay work which is done thus Pitts are sunk upon the Lowest Level of the coal which Can be conveniently come at which is done that all the Coal which is got may be Drawn down hill to the pits from whence it is raisd They then drive narrow passages from pit to pit & across the dip of the Coal in as many places as are nescessary upon a dead Level these are calld Ends & divide the coal into what is Calld ranks which are divisions of as much in the Breadth as can easily be got These communicate with each other at the Extremities by drifts drove up hill Calld narrow Bays.

These outlines being finishd they begin to get their Coals by entering stalls or bays always beginning at the upper rank which as farthest from the pitts in use they mean to Leave first to tumble in & destroy itself These stalls or bays are done by beginning to work uphill with narrow Entrances widning their work as they go in till they have drove almost through the rank supporting the roof behind them with Timber when they have drove as near to the Last End as they chuse they begin to return robbing their pillars as it is termd (that is taking them away) & still taking the wood that is behind them till they again come near the End where they began there they always Leave Large & substantial pillars to support it as a passage for air which they may always return to in this manner they proceed till they have got all the coal between the new & the old pitts when it becomes nescessary to sink new ones as drawing uphill is never to be done except in cases of nescessity.

In the Course of this day I had many opportunities of examining what are Calld faults in coal which appear to me to proceed from the subsidence of strata pressing & carrying down below their proper places anything under them as where one part of a stratum of coal is thrown much below the other so that a person driving in a coal instead of meeting with a continued stratum of it meets something Else as either the coal he is in is sunk below that he expected to meet or that he expected to have met is sunk below that which he is in in this Case the experiencd miner Knows his remedy instantly first by a rule which never fails & is that if his Coal fails him at the top he expects that what he searches for is subsidied & vice versa but if as sometimes but very rarely it happens that the Coal is broke off quite Even he then Examines the Stratum which he finds instead of it if it answers in thickness & sort to any he found in driving his pit down to the coal he concludes that his coal is gone down if not that it has gone up (as they term it) that is that the coal he is working in is gone down

One thing is observable here which the Colliers Call Slip Joints these are seen in the root of the coal but never unless where a fault is there the rock is divided into many small plates whose sides are by no means flat but are polishd in a very high degree which probably was done by their rubbing against one another at the time of the subsidence of the strata

[30 December 1767] Went to Manchester but could not get time to see so good a town to the advantage I could have wishd the streets however seem in general well Built some of them remarkably so but the architecture of the Publick buildings does not suit my taste so well the Exchange[21] particularly & the new church[22] the first seems heavy to a degree the second is a mixture of Gothick & grecian Architecture which possibly may please some tastes but mine I confess it does not

Here is also a publick library[23] which I saw Endowd by a Mr Cheatham about 1564 with 150 pounds a year to buy books it Contains a Large Collection & many good ones here is also a small Collection of rarities hung up museum wise

[31 December 1767] This day Finishd viewing the Dukes works underground by viewing his Level & his machine for raising water

What is Calld the main Level or that at which you first Enter the rock is already driven very near a mile in Lengh & very much more is intended it goes directly against the Coal being intended to dry it From this branches another which is calld the side Level this is already driven about 1000 yards & is so wide that two boats may pass in any Part of it this also goes forwards & no end is yet fixd for it so that there appears to be no end to these works unless some strange ill fortune should throw in accidents at present not in the Least expected

In the Lowest part of these works the duke is obligd to lift water which has given opportunity for shewing an Engine for that purpose of an intirely new construction whose very powers & principles have never been before thought of the weight of a column of water thrown down a shaft in wood pipes is here made to lift a piston which works 2 lifts of Pumps with Clack to take of & Lay on the Power much in the manner of a fire Engine more I should say of it but that I insert a plan of it [not reproduced here] drawn by its Inventor Mr Ashton Tonge who is Employd in the Dukes works

It remains now that I should give some account of the Canal in General the whole intended Length of which from Worsley to Liverpool including the Branch to Manchester is 34 miles of which 21 are Compleated & 13 Remain unfinishd The Expence incurrd by this undertaking must be immensely Large when we consider that the Cutting a Canal navigable for vessells of 40 tuns & upwards is not all the Canal is to be carried over rivers some of which are navigable & this upon arches of immense strength to support so large a weight as well as vast hight to admit Large vessels to sail under them without difficulty as is the case

The benefits accrueing to the Countrey are are also almost invaluable trade is opend between two very Large Towns before Labouring under great inconveniences The price of Carriage between them tho there was before a navigation[24] is Lessend & a plan is struck out before deem'd impracticable which has already been followd in several parts of the Kingdom in a more intensive way at publick expence & is likely in time to be of the most essential use to the nation as it will open inland Commerce in time of war without its being Exposd to the dangers of Privateers by which our Coasting Trade has always sufferd very materialy

These are advantages which occur at the First Sight but what may we not Expect from such a plan followd up with spirit in a countrey depending intirely upon her Commerce as this does for all however which we do reap from it we must acknowledge ourselves undebted to its noble author & not a little to his cheif executor Mr. Jno Gilbert whose most indefatigable industry himself overlooking every part & trusting scarce the smallest thing to be done except under his own eye. . . .

I shall take this opportunity of confuting a notion which has a good deal prevaild from its seeming reasonableness & has prejudiced canals of this Kind in the opinion of People who have been misled by it I mean that of Canals being sooner Frozen over & navigation sooner stopped upon them than on rivers this would undoubtedly be so were things Left to themselves from the streams of the Latter which Keep open Part of them at Least throughout the Course of Long & Sharp frost but their wider parts where the stream moves Slower are frozen almost as soon as the Stagnant Canal these from their Breadth it is scarce possible to Keep open even at first whilst in the narrow Canal the Ice is Easily Broke & never sufferd to come to any thickness till a Long Continuance of a Keen Frost which has since I have been here Froze to bear the weight of a man between 11 at night & 8 in the morning The navigation of the Irwel was this time stoppd two days before the Canal gave over I myself saw nine barges the day it gave out a little above Barton Bridge attempting with all their Powers of men & horses to get up but in vain whether made more Eager to accomplish their Endeavours by seeing the canal navigation goin on without interuption I cannot tell but one of them was drawn against the Ice till she had a hole cut in her bottom which would have sunk her had she not been immediately unladen

Many of the Common things which were Found nescessary in this canal are struck out upon new principles & much improve upon ones that have been thought of before the wastes[25] for instance are made by round pools beyond the Hawling road[26] which communicate with the Canal by an arch Layd under it (Figure 10) in the midst of this is a

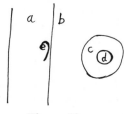

Figure 10.

a canal   b Hawling Rd   c waste   d round hole   e arch

round hole through which the water falls onto the brook prepard to receive it the merits of this are 1st that it can never be liable to obstructions on account of the arch being Layd under water 2cnd the Hawling road is free & undamaged tho Wastes are often nescessary on the same side the water as it is 3d in case anything should happen to the waste its communication with the canal is immediately & easily stoppd by any Kind of plank put against the arch on the inside which being held in their place by the water make a substantial dam till the whole is repaird

At Castle Feild in Manchester the Canal[27] is driven underground in a tunnel upon this a pit is sunk by which means coals are drawn up to the top of a hill & Loaded into carts which Can Go & come so much the oftener as they have not the hill to go up

This is done by a crane which works by water on the same principles as mill tackles or indeed as turning frames go by which is called the Endless rope. . . .

The Lowermost wheel being turnd by a waterwheel turns the rope & the upper wheel by which the goods are lifted up this is Let out of the Geers by a wind at the top which has a rope fastend to the upper wheel by drawing this tight the upper wheel is raisd & of course the rope which goes round them both is made tight & all work on the Contrary by slackening the wind the upper wheel is suffered to drop Lower the Endless rope of course Slackend & the upper wheel no longer moved

These & many more usefull & ingenious inventions were thought of and executed by Mr. Brindley[28] who also did most of the Engineering work of the Canal he is a man of no Education but of Extremely strong natural Parts he was recommended to the duke by Mr Gilbert who found him in Staffordshire where he was only famous for being the Best Mill wright in the Countrey

[2 January 1768] Left Worsley this day about noon at night came to Northwych 22 miles went immediately to see the salt works[29] went down into their Largest pit but was so unlucky as to find the air in it so bad that it was with difficulty that I could Keep in a candle could I have had Light Enough the sight would have been one of the most beautifull & magnificent that Could be Conceivd imagine an immense vault Covering 3 acres of Ground 48 feet high supported by pillars 24 feet in Diameter the whole made of a material which tho unpolished Glitters more than the most Polishd Marble for its sides like so many Chrystalls throw Light a thousand different ways

The Rock of Salt which in this & many more almost as Large Caverns are Hollowd is a most extrodinary Production of nature it has been tryd for about 2 miles in Lengh & at the edges of that is so thin as scarce to be worth working tho in the middle its depth has never been tryd as they get full as much as they want at small depth but I was told that it was bord 50 yards without any Bottom being found it is unlike all other rocks that Ever came under my observation in that it is perfectly solid & free from all Joints & cracks its roof however is very uneven Causd perhaps by water which has melted holes in it which are supplyd by the stratum above that is not Liable to be melted

Tho Calld Rock it is absolute Salt to all intents & purposes some perfectly pure & clean in appearance much Like Spar but the Larger Part Tingd with some Kind of Dross which gives it a reddish Colour in it are some Lumps of Stone inclosd which the workmen Call Flints it seems to be of a Brownish Grit but I was not Lucky Enough to get a good Peice of it tho the workmen told me they sometimes found Peices as big as their Heads

They get it down with Picks made very strong & Heavy with these they bring up stepps about 1 foot & $\frac{1}{2}$ or 2 feet high Pulling it down as they go by striking behind it if it fails of Coming Easily pouring a little Brine into the nitches they have made which is sure to bring it down with a very little patience sometimes in Peices 2 or 3 tons weight. . . .

This day went to see a new invention to quicker spinning set up in this Town [Cheadle, Cheshire] by Geo: Goodwin who had it from a Freind of his in Yorkshire it is Certainly very Ingenious & simple but so difficult to Describe Intelligibly that I have Sketchd out a small plan of it & Letterd Every part of it so shall only mention its Effects it is said by its master to do work 4 times Easier & consequently as much Cheaper than the single wheel it has also another advantage for the room being made of a proper Lengh the threads are never reeld but immediately warpd upon pinns fixd upon the side of the room for that Purpose & in that Shape is sent to the Bleaching Crofts & from thence given immediately to the weavers & it is reckond that two spinners will Keep three weavers at work

They have not yet made any Cloth of above half a crown a yard but as the invention is quite in its infancy it will probably be brought to much greater perfection & that before Long for tho it is at Present only in the hands of two people it is more than probable from its Extreme simplicity that it will soon become common[30]. . . .

[12 January 1768] Came today through Exceeding bad roads to Watling Street upon the Famous roman Road of that name Here [at Coalbrookdale] I resolvd to stay some time to view carefully the Famous Iron works in this neighbourhood belonging to the Quakers Company this Took me up two days but for the sake of varying my observations into some order I shall Join them together

These works are situate between the Severn & the Great turnpike road between which Points they have a rail way Layd for the Convenience of their waggons their Cheif Erections are at three different Places Ketley Horse hay & Coalbrookdale in which they work 7 furnaces & a Forge & turn out a vast Quantity of Iron Mr Reynolds[31] a Quaker who seemd Particularly Careful of his Speech the Cheif manager told me a great deal more than 100 ton a week I should think more than 150 as their Furnaces I am told have yeilded as much as 30 & are seldom under twenty

First then Ketley is Situate Just by the turnpike road side here there are three Furnaces which are supplyd with water by two Fire Engines[32] The Largest of which is 68 inches Cylinder The Furnaces are Each of them 12 feet wide & about 31 in hight in the Form of a double Cone blunt at Each End here I saw a Cast of Piggs which was done thus

The Casting room From the mouth of the Furnace to the wall was about 48 feet Long the floor of this is Covered with sand For about two yards broad which is Laid Close up to the side of what is Calld the Sow a Long bar of Cast iron reaching the whole Lengh of the house along the side of this is made a groove which is Calld the runner & From the side of this are cut a number of short ones every one about a yard in Length which are to be the Piggs all which is done with a wooden tool like a hough[33] with a round Edge The Iron is then Tappd at the End of the runner which is not a very Easy work tho it is shut in by nothing but wet sand thrown in with a shovel as this by the action of the intense Fire is Partly vitrified it is nescessary to do it with the Point of an Iron Crow which is hammerd in by one man while another holds it till it touches the Liquid Iron which immediately Follows almost as Fluid as water & running the whole Lengh of the Runner Fills it & all the Pig moulds which Communicate with it The Cast which I saw was 54 but they sometimes Let it run as far as 60 [cwt]

While the Piggs are Casting the Bellows are suffered to go but gently but the Small wind they occasion finding vent at the hole in the Furnace made for the dross to run over fills the room almost intirely full of sparks making a most beautifull appearance The waste of the Cinder [when tapping the slag] before Casting Affords an appearance which Gives the Idea of the rivers of Lava running down the sides of a volcano in an irruption streams of Liquid fire issuing out From thence & dispersing different ways still run slower as they become cooler

From hence to Horse hay is two miles here are two furnaces which are supplyd with water by one fire Engine for tho there are two here yet only one works the other standing by to releive it in case of accidents

These five Furnaces are Employd intirely in Casting of Piggs a great Part of which go to the Forge to be made malleable Iron tho some are melted again in air furnaces for castings

To Coalbrookdale is from hence two miles here are two more Furnaces Employd intirely in the casting way which Probably Cast Greater Quantities & better metal than any other in the Kingdom here all Large Casting work is done in the Greatest Perfection they often cast Cylinders for fire Engines as Large as 72 inches diameter & have cast one as high as 75 I should have thought myself extreemly fortunate to have seen any of Their Large ware Cast but was Forcd to content myself with a 90 Gallon Pot the Largest they then had

moulded The moulds for these & all smaller Potts are made of sand
and charcole dust moulded in a wooden Frame into these they Pour the
melted Iron From Ladles made of Iron & coverd with Clay to Preserve
them from melting as soon as the mould is Filld before the Iron is
Compleatly set The air that was Containd in the hollow of the mould
rarified by Extreme heat seeks for vent causing a report sometimes as
Loud as a pistol but without doing the Least damage

The moulds for Larger vessels I saw making tho none were ready to
Fill they are made of Loam Laid on upon Frames of Brick & dryd with
the Greatest Caution For if the Least Damp should remain it would
not only destroy the Pot but would cause the metal to Fly about the
melting house like shot to the infinite danger of Burning Every body in
it

After this short account of their works it will be nescessary to say
something of their method of making Iron in doing it they use 7 Kinds
of stone all Found upon or near the spot Blackstone Blue Flat White
Balls white Flat Pitchy Loggs Pennystone & yellowstone Earth of these
the two First are reckond the Best yeilding stones & tho they are all
workd together almost twice the quantity of them is put than of any
other sort next to them the Pitchy Loggs are Esteemd which are a kind
of iron stone abounding with much Liquid Bitumen which sticks upon
them Like Pitch the rest are Pretty near Equal but the white Balls are
very Peculiar they are nodules sometimes very Large other times not
Larger than a Walnut upon being broke now & then one shews the
appearance of a Plant Generally either Fern or Palm one side shewing
the relief the other the Cast or mould of it

All these stones require to be Calcind before they are put into the
Furnace this is done in the yard by Laying them upon Floors of Coals
Layd about 6 inches thick these are set Fire to on the windward side &
are soon sufficiently Calcind They are then Put into the Furnace which
is supplyd intirely with Coaks (not a bit of Charcoal being used in any
of these works) & about one third of their Quantity of Limestone Put
with them as a Flux in which they Gradually descend as Lime does in
a Kiln only with this difference that the Fire grows hotter still as they
descend Lower two Pair of monstrous Bellows Blowing in at a small
hole made but a little above the Bottom

having now Traced the Iron through the Furnace we will attend it
to the Forge where it is to be made into malleable Iron of these the
Company have but one at the dale they selling much Iron in Piggs &
having one or more besides at Bridgenorth in this however Iron is
made in a new manner for which they have lately got a Patent[34] it is
done intirely with Coaks by this method two Air Furnaces are
Constructed by which is meant small ones little Bigger than Large
Ovens made of one single arch on one side of which is a grate to make
a large Fire on the other a chimney to draw that fire so the Flame
beating very strong across the intermediate space Causes an intense
heat there into which the Metal is Put in the First of these a fire is

made Just sufficient to melt the Piggs which are put into it which as soon as they are in Fusion or scarce advancd quite so far as taken out from thence much in the appearance of half vitrified Cinders they are then removd into the other where as intense a fire as Possible is made here they come to nature as the Forge men term it that is the metal Casts off the dross & runns together in a state fit to be immediately put under the Hammer where by six or seven strokes it is made into an octagonal Bar very rough but fit after this to be Drawn out into Barrs by a very moderate heat

The Hammer by which this is done is a Large Block of Cast Metal fixd on a Large beam of wood This is moved by water & Every time it is raisd strikes against another Beam Layd Paralel with it calld a rabbit from whence it recoils with the force of a spring

The Iron made by this method is just as good in its Qualities as that made with Charcoal but there a Large waste of it in the intense heat of the Second Furnace which makes their Invention not near so profitable as was Expected[35]

All the Iron made here Probably From the Quality of the stone is of the cold short Kind (the name given to that which is brittle when cold) sometimes it inclines to the red short (as it is calld when it will bear the hammer when white but is brittle when it becomes red hot) this is the worst Quality Iron can Possibly have & they try all in their Power to correct it by Proper mixtures of the different stones

I must now say a word or two of their rail way as it is the most extensive one I beleive in this Part of the Kingdom & the waggons upon it the Best constructed

it is made with two Frames the side timbers of Each 4 inches square those of the Bottom Frame 9 Joind together by Cross Timbers Calld Sleepers which stand about 5 in two yards & Keep the side timbers steady to support the upper ones which are pinnd Lenghways upon them with wooden Pinns (Figure 11) these are made of the

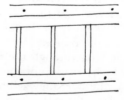

Figure 11.

Firmest heart of Oak that Can be Got & even that wears out very soon by the immense weights of the Waggons so much so that they have began at the dale to make the upper barrs of Cast Iron & have thought of Continuing it all their ways

The waggons themselves are made prodigiously strong their Lengh

about 10 feet Breadth 4 the weight of Each of them about 22 hundred the axletrees are Cast Iron & move with the wheels which are Cast iron likewise the inner Edge of them overhanging about an inch to Keep them upon the ways the proper Load of one of them is $2\frac{1}{2}$ tons but they will sometimes carry 4 nay 5 ton with them in which Cases they draw with five horses tho their Proper number is only three they have many hills in the road some of them steep these they go by this Contrivance (Figure 12) they have a peice of wood cut in the segment

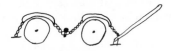

Figure 12.

of a circle Calld a Break to fit Each wheel these are fastend together by Chains one End of which fastend to a staple before the forewheel the middle Passes under a roller between the wheels & the other End hooks upon the End of a pole 12 feet long Calld the Jig pole which rests in a staple fastend on the hind Part of the Carriage & acting as a lever is directed by the hand of the driver to Confine the breaks Close to the wheels causing just as much Friction as is nescessary to make the waggon go at a proper degree of Slowness in case an accident should happen notwithstanding this caution which Can arise from nothing but the Carelessness of the driver there are peices of wood set upon pins which are Easily swung across the road diagonaly these catch the wheels from their proper tracks & immediately carry the Carriage against a Little sloping Bank made for the purpose where it immediately stops. . . .

[21 January 1768] Went this morn to a place Calld Soho about two miles from the town where Mr Boulton to whom I had a Letter Lived & Carried on his Manufacture which he does in a very noble way his People all working in a very Large Building the architecture of which is good

Here he makes Cheifly Buttons toys in steel & tortoise shell watch Keys &c to describe the different operations performed upon a Button would be too tedious sufice it to say that after the Plates were rolld they were shaped pressed & cut to proper sizes in dies just the same as Coining is Performd in after this they are Burnishd by a tool made of Bloodstone [heliotrope] upon a turning frame after which they are fit to file The moulds upon which they are put are also madde in a Lathe with vast quickness & dexterity every one being cut off by a sharp tool at one operation these are afterwards drilld by a Lathe which turns 4 spindles at once then threads of Catgut are tied through the holes & the Button being filld with Melted rosin these are applied upon it which Finishes the manufacture

Here are also another sort of Buttons made which are inlayd with steel these are first Cast in Copper then a die being prepard in which small peices of steel as the Pattern requires are Laid the Button made red hot is Laid upon them then a weight being Let down at one blow fixes all the Peices of Steel into their Places & at the same time makes the Shank

Here are also many watch chains & Keys made which are put into form by making the parts red hot & putting them under a die the fine Ends of the Keys are afterwards turnd upon a frame open work they have often occasion to do in that Case every hole is made by a distinct tool so that a Button in that operation only often goes through 5 or 6 hands

Steel is here polishd to great perfection Equal at Least to Woodstock the most difficult part of this manufacture is Procuring iron free from Flaws or specks small quantities of this is generaly Procurd by watching when old wheel tyre is Broke But Mr Boulton tells me that he has invented a Kind of Iron homogenious in its Parts and consequently free from all such imperfections which he has began upon & hopes in a little time to procure Enough for all his works

The polishing is done at many operations First upon Lapidaries mills then Brushes are Employd but the Last Polish of all fine steel work is put on by the hand that is by rubbing on I believe Putty with the Bare Hand.

The works of Tortoise shell I did not see going forward can therefore only tell their Effects they have methods of Clouding the inside of the Shell with many Colours both opake & Clear & some very beautifull in one case they had introduced two wings of a fly & a peice of a feather I mean the natural substances they have also the art of making a composition of the shavings which is very beautifull & solid. . . .

Notes

1. Cf. H.C. Cameron, *Sir Joseph Banks* (1952).

2. A.M. Lysaght, *Joseph Banks in Newfoundland and Labrador 1766* (1973).

3. S.G. Perceval (ed.), 'Journal of an excursion to Eastbury and Bristol etc in May and June 1767', *Proc. Bristol Nat. Hist. Soc.*, IX (1898), 7–25.

4. MS Add 6294. I am grateful to the Librarian of the University of Cambridge for permission to publish portions of this journal, and to Mr F. Stitt, Staffordshire County Archivist, who made a microfilm of it available to me.

5. See *History of Technology* (1979), 107, fig. 3 (Ed.).

6. The manufacture of white lead (ceruse) by the action of vinegar on lead was first described by Theophrastus in the third century B.C., and subsequently by later classical, medieval (Theophilus Presbyter, c. 1140) and Renaissance writers.

7. In 1698 it was reckoned that dissolved rock salt could be refined with coal at 5s.6d. a ton on the Mersey estuary near the pits, whereas brine in Cheshire had to be boiled with coal at 16s.8d. a ton because of its transport cost (W. Chaloner in *Trans. Lanc. Cheshire Antiquarian Society*, **71**, 1961).

8. It had previously been described by W. Efford in *Gentleman's Magazine*, August 1767.

9. Presumably John Gilbert (1724–95), the Duke of Bridgewater's agent who was in charge of his canals. He was acquainted with James Brindley.

10. The fine Staffordshire salt-glazed stoneware had been perfected by John Astbury (1688?–1743). It was close to porcelain in translucency and hardness. He also began the practice of putting ground flints into the body (Ed.).

11. Presumably what is usually called 'creamware', which several potters had begun to produce, Josiah Wedgwood (of Burslem) among them. Wedgwood developed his 'Queen's Ware' from this type, which was lead-glazed and used the white-burning clays of Devon and Dorset (Ed.).

12. Saggers.

13. One such flint-mill is preserved at Cheddleton, near Leek (Staffs.).

14. Brindley's 'Grand Trunk Canal' now known as the Trent and Mersey. The Harecastle Tunnel took eleven years to drive; it was only 6 feet high by 7 feet wide and boats had to be 'legged' its whole length of 2880 yards. Telford cut a new tunnel in the nineteenth century and in time the original became impassable through subsidence.

15. Begun in 1761, it was intended to bring coal from the Duke's mines at Worsley to Manchester. The canal was later extended to Runcorn on the Mersey.

16. Dunham Massey, $2\frac{1}{2}$ miles west of Altrincham.

17. John Kay's flying-shuttle patented in 1733.

18. Opposition to Kay's innovation drove him to France in 1747; he returned to England in 1759, but lived in France again from 1762 to 1765. During his residence there he received a pension from the French crown (W. Chaloner in *Bull. J. Rylands Library*, **48** (1), 1965).

19. For a general account of the English alum industry see Charles Singer, *The Earliest Chemical Industry*, (London, 1948), Chs. 7 and 8 (no mention of Blackburn).

20. At Worsley the Duke's Canal enters two low tunnels in a rock face and continues underground to the coal seams. The tunnels have not carried coal since 1887 but still serve to drain the mine.

21. Built 1729, replaced 1790, an open structure of Palladian style.

22. Probably St. John's, Deansgate.

23. Chetham's Library (1656), founded by Humphrey Chetham with an endowment of £1,000, is still in existence. Karl Marx and Friedrich Engels used sometimes to meet here when the latter was employed in Manchester.

24. The Mersey and Irwell, opened in 1726, but spoiled by neglect.

25. Overflow weirs to keep the pound level constant. 'Well-weirs' exist on Brindley's Staffordshire and Worcestershire Canal as by-pass weirs to locks but are not general on the Bridgewater Canal.

26. Towpath.

27. Remains of the works described here were found in 1960. The waterway was led into the side of the Castlefield bluff by means of a tunnel, to which a vertical shaft penetrated from the surface above. A swivel crane was erected over the shaft to lift 8cwt boxes of coal from the canal-boats, which had been loaded at Worsley. Water drawn from the River Medlock powered the crane.

28. James Brindley (1716–62), archetypal engineer of the English industrial revolution canals.

29. The rock-salt deposits at Northwich had been discovered in 1670 when borings for coal were made.

30. This prediction was not fulfilled.

31. Richard Reynolds (1735–1816) was son-in-law to Abraham Darby II; he managed the Coalbrookdale works until 1768 when he handed over to Abraham Darby III (1750–89) though retaining the Ketley part of the enterprise under his own direction.

32. Newcomen engines used to pump the spent water in the tail-pond back above the wheel driving the furnace-bellows. These engines were introduced in 1754.

33. Hoe.

34. Patent no. 851, June 1766. The new process, using coal (not coke, as Banks has it) was devised by the brothers Cranage. The first heating reduced silicon and sulphur in the iron; the second heating reduced its carbon.

35. Until the method of 'wet-puddling' on an iron-oxide bed was introduced in the 1830s all methods of producing wrought iron from pig entailed relatively large losses, since the combustion that reduced the carbon also oxidized the iron, which was lost in scale and slag.

Water, Stampers and Paper in the Auvergne: A Medieval Tradition

RICHARD L. HILLS

*As part of the National Paper Museum, the North-Western Museum of Science &
Industry received a water-powered stamper of the type used in small mills form
preparing paper-pulp. In 1977 Dr Hills visited the neighbourhood of Ambert (Puy
de Dôme) in France in order to examine the small paper-mills still surviving in
that region (see Map 1). Ambert is a town deep in the Massif Central, about
equidistant (130 km) from Lyon and Clermont Ferrand. The stamper in
Manchester corresponded very closely in dimensions to the one measured at Ambert,
and a reconstruction based on the Auvergne mills has been created in the
North-Western Museum.*

Ambert

There is a claim that the paper-mill of Richard-de-Bas at Ambert was
in existence by 1326 (Map 2). Whatever the truth may be,[1] it is certain
that the provincial capital of Ambert was one of the earlier
papermaking centres in France. One reason is not hard to seek, for the
streams that flow down the small valleys from the mountains on either
side provide not only clean water for the papermaking processes but
also power to drive the machines. It has been claimed that there were
over three hundred mills in this vicinity,[2] not all of them paper-mills of
course, but this number suggests that Ambert was at one time an
industrial centre of considerable importance. What remains may,
therefore, tell us a great deal about the type and scale of medieval
enterprises.

The main papermaking area close to Ambert, where the industry
has survived to the present day, is along the streams Gourre and Lagat
flowing from the mountains to the east, past the mills and through the
town of Ambert itself before joining La Dore, the main river of the
region (Map 3). The mountains reach nearly 1,400 metres in height
and the streams appear in springs at about the 1,000-metre level, but
at that height are very small. Ambert is about 500 metres high. On the
Gourre, there may have been up to eighteen waterwheels before it

143

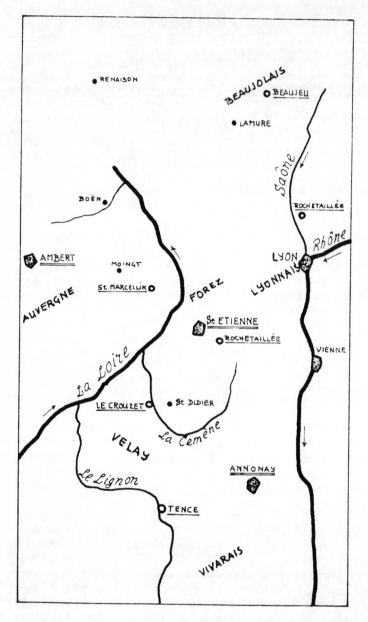

BEAUJOLAIS-VIVARAIS

Map 1.

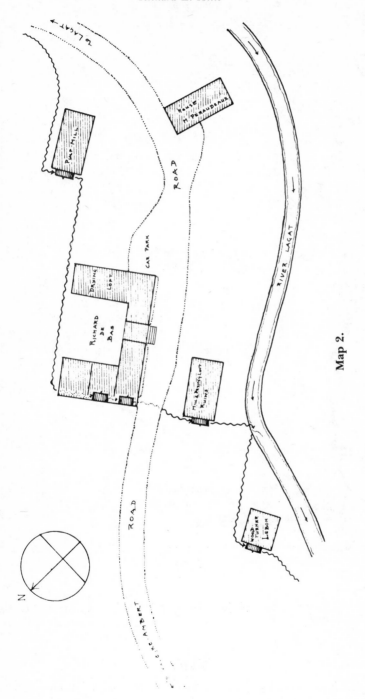

Map 2.

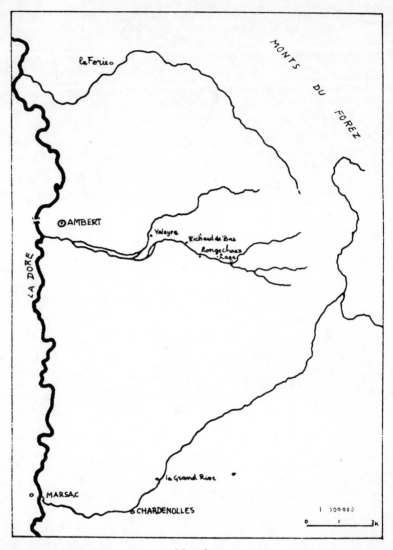

Map 3.

joined the Lagat, while in this valley another fifteen wheel sites have been identified.

Water

With so many mills, a visitor from England would have expected to find extensive reservoirs, dams, weirs and sluices to control the water, but there is nothing. Either work in the mills was seasonal (which it is not today), or the water flowed freely all the year round at roughly the same rate all the time. There is no evidence that the work was ever seasonal and it is surmised that the location was exceedingly favoured with regular rainfall and with mountains which act as vast sponges, soaking up the rain and letting it out slowly through the springs.

In the Lagat valley the stream is diverted into a channel (or *bief*) along the hillside into the upper part of Lagat village itself (Map 4). It first drives Begonin's mill which is still working (1977) with its machine, then past La Lagat mill with its wheel now idle, through another abandoned mill and then on to the two mills of Lebon, both now alas defunct. At the point where this tailrace flows into the Lagat stream again, it is tapped into another *bief* and diverted along the side of the valley to an unidentified mill and so to Fournier's mill where the wheel is still used to generate electricity, but the papermaking machine was stopped (temporarily one hopes) through the illness of M. Fournier.

Fournier's mill stands by the main stream so the tailrace here flows back into the Lagat again, but it is immediately tapped to supply the wheel of the Richard-de-Bas pulp mill, the two wheels in the Richard-de-Bas mill itself and one, or possibly two mills, between the Richard-de-Bas mill and the valley bottom (Map 2). Similar instances of this multiple use of the same water could be quoted for other groups of mills in this valley and in the Gourre valley as well. If a mill is idle, the water passes down the overflow to the next one, while if the water is being used to drive the mill wheel, it flows round the wheel and on to the next without any hindrance either. There is no attempt at any conservation or storage of the water; what power is not used is wasted.

The levels of all the *biefs* have been carefully determined to make the maximum use of the fall in the streams. In the Lagat valley particularly, the *biefs* and mills follow so closely together that very few metres of fall in the water are unexploited, although the wheels may vary in diameter. One might have expected to find traces of older, disused *biefs* in these groups of mills bringing the water in at a lower level to say the bottom mill only, and that the higher mills would have been built later with a new channel. This would have seemed particularly appropriate for the Richard-de-Bas mill which is situated quite high up on the side of the hill. No traces of alternative *biefs* are left today, so it must be assumed that the original owner foresaw the development of an industrial estate on his land and laid out the water channels accordingly.

An instance of this may be the highest mill in the Gourre valley at La Rodarie. The stream here seems too small to drive any mill, but a map of 1863 preserved in the Ambert Hôtel de Ville shows that the present farm set high on the hillside on the road to Banquebout was a paper-mill. The *bief* can be traced by the road and the water must have flowed back to the Gourre down the side of the hill, past a small

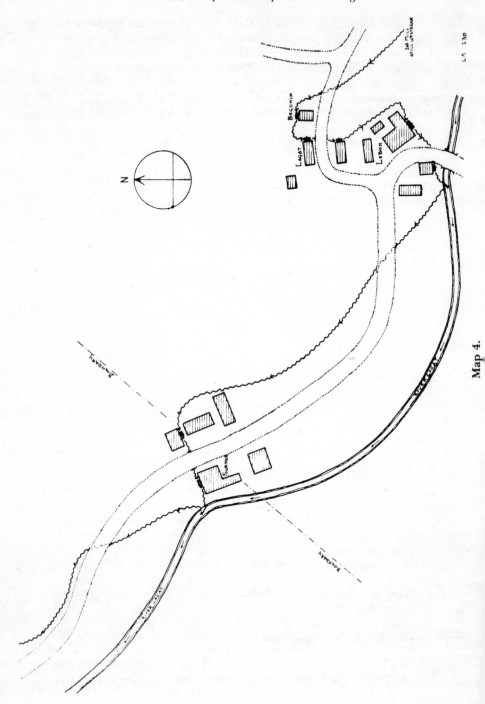

Map 4.

corn-mill. This mill has a single set of stones and the remains of the waterwheel rotting away outside, so even up here sufficient power could be derived from the stream. If there had been sufficient demand, one or two more mills could have been built on the side of the hill between the paper-mill and the corn-mill.

Stampers

The next question that must be asked is why there were so many little mills with wheels of about 3 metres in diameter and 80 cm broad and not one large wheel driving a single large mill. The answer seems to lie in the scale and development of medieval technology, and particularly of papermaking technology. The only method of making pulp at that time was with stampers. At the Richard-de-Bas lower mill, the complete set of six stamper troughs with three hammers per trough still survives (one set of hammers has been removed to allow access through a new doorway). A similar set was recently scrapped at the Lagat mill and one of the troughs with its hammers has been preserved in Basel and possibly another in Manchester. The set of six stampers with vat and press on the ground floor, living house on the next and drying loft on top probably composed an economic unit which, given the technology of the times, was difficult to expand except by duplicating everything.

The crux of this assessment lies in the stamper unit. The waterwheel, at first of undershot type,[3] would have been comparatively inefficient. It was mounted on one end of a long wooden shaft made from the trunk of a single tree. To one side of the wheel the shaft was supported by an end-bearing while, on the other side, the shaft passed through the wall of the mill to the stamper chamber. A few moments inside the stamper chamber with the stampers at work makes the visitor realize why these rooms were encased with solid stone walls and vaulted roof to prevent most of the noise reaching the rest of the mill. The shaft with the cams fitted to it passed behind all six stamper troughs to a bearing at its other end. There used to be no intermediate bearings, but today pairs of rubber-tyred wheels have been placed in the middle to give additional support. It is suggested that this shaft, about 9 metres long, is the limitation in the design and development of these mills. At some period the waterwheels have been made overshot, which must have necessitated a change in the layout of the mills, for the stampers must have been turned round the other way in their buildings to suit the rotating of the cam shaft now in the opposite direction, but the buildings do not appear to have been lengthened to take advantage of the extra power.

What were the alternative ways of driving more stampers? One method would have been to extend the shaft by devising a bearing and a coupling to join two shafts end to end. In this way more stamper troughs could have been introduced in the same line, but again there is no evidence that this was ever done. The problems of taking the torque from one shaft to another may have been too great, except through gearing. An obvious solution would have been to use tail hammers where the hammer is raised by pressing down on the end of the helve remote from the head, but this never seems to have been employed in papermaking at the pulp stage.

Two sets of stampers were sometimes driven off a single cam shaft by employing an intermediate linkage to operate the second set. The cam shaft lifted one set of hammers in the usual way and on its other side pressed down one end of a link pivoted in its middle. The far end of this link lifted the hammer and so operated the stampers. This is a simple but clumsy solution and not one that met with any lasting favour.

Another method was to introduce gearing. A common gearwheel on the waterwheel shaft drove two parallel cam shafts so that in this layout the stampers all faced the same way. With the waterwheel fixed on the end of one camshaft and the second driven off it, the stampers faced each other.[4] With a direct drive, there were no losses of power through the transmission. Early forms of gears were notoriously inefficient, and there was also the problem of driving a machine which needed intermittent pulses of power through relatively weak wooden gearing. In an area like Ambert with a water supply that was comparatively small in volume but steady, it must have seemed more sensible to build small mills where each was a complete unit by itself, and avoided these other problems.

One other way of increasing production was to build a double mill. Richard-de-Bas is an example of this (Map 2). Originally there must have been two sets of everything, with one set placed higher up the hill but under the same roof so that the water first drove the higher mill wheel and then passed on to the lower. Traces of a similar arrangement can be found at Nouara on the Gourre where the water was first taken to a building now demolished, then over two wheels behind the building now converted into a hostel and then over a fourth wheel, possibly a corn-mill at the bottom. The Grand Rive mill to the south of Ambert in the Grandrif valley was the largest mill in the region.[5] In 1676 there would appear to have been seven wheels with thirty-eight sets of stampers and four vats, but by 1864 these had been reduced to three wheels only and two vats. The original water channel and the sites of these wheels can still be seen (Map 5) and in addition here is the only example of the installation of a higher *bief* for, probably when the building was converted into a brewery, a second higher channel was built to supply a turbine through an iron pipe.

The stampers at Ambert raise the question of the accuracy of the seventeenth- and eighteenth-century pictures of papermaking. All that arrived in Manchester of the stampers now in the National Paper Museum Collection was the stone trough, three hammers, the wooden

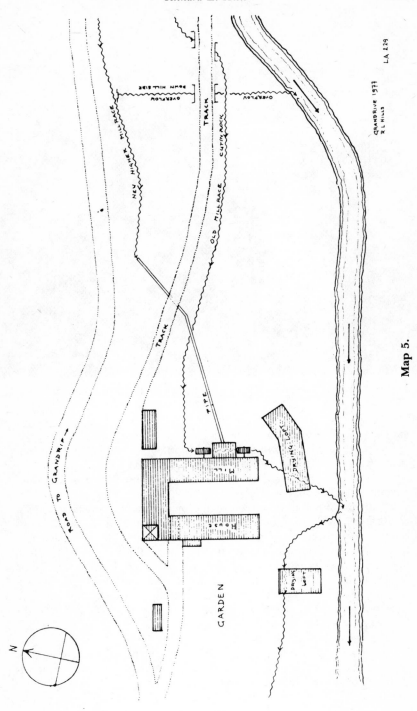

Map 5.

Figure 1. Reconstruction of the Ambert stamps in the North Western Museum of Science & Industry, Manchester.

support for the pivots and the guides for the hammer shafts. These pieces had to be re-erected safely and, if possible, a cam shaft reconstructed to show how the stampers were driven. What better source of assistance could be found than the illustrations in La Lande, Livourne, and similar works?[6]

With the possible exception of the picture in V. Zonca, *Novo Teatro di Machine et Edificii* (1607) which it is very difficult to puzzle out in order to determine what goes where, the shafts of the hammers in all the other illustrations appear to lie horizontally, with the heads vertical, when the hammers are resting on the bedplates in the bottoms of the troughs. But the stampers in Manchester could not be fitted together in that way. A closer examination revealed that the heads were not at right angles to their shafts and that the iron spikes in their feet were mounted at such an angle that the shaft must have been sloping upwards from the pivots.

A visit to Ambert confirmed this arrangement, and a few moments playing about at the geometry with pencil and paper will reveal why. With a horizontal shaft and the stamper head set at right angles to it, the small lift of about 6 inches (15 cm) will mean that the hammer will

Figures 2a and 2b. Ambert, Richard de Bas Mill. Two views of the stamps which are in a basement with a vaulted roof. The camshaft from the water-wheel runs alone the far wall; one set of three stamps is in operation, the others being lifted clear by the hooks at their tail ends.

Figure 3. Ambert, La Lagat Mill. Note the ventilated drying loft at the upper level, and the water-wheel in the channel at the side.

move only vertically up and down. While this will give a good pounding action, there will be very little stirring or circulation of the pulp. If the positions of either the pivots or the cam shaft are altered so that the shafts are inclined, then the stamper head will have a horizontal component as well as a vertical one in its movement. Further, if the heads themselves are inclined on their shafts, they will give a greater stirring action when they rise and fall.

A survey of the stampers remaining in their original situations[7] shows that they have these features. At Ambert, the shafts are inclined upwards from the pivots to the cam shaft and the heads are mounted just off a right angle on them. This would appear to have been the normal practice in central and northern Europe. At Capellades, the hammer heads are set at a very pronounced angle to their shafts. On the other hand, at Pescia and Fabriano, not only are the heads steeply angled, but the shafts slope downwards from the pivots to the cam shaft which is placed very low down. The questions that now have to be asked are when these features were introduced and how accurate are the early illustrations.

There is a further detail about the stampers in both the Richard-de-Bas and Lagat mills. The hammers with the coarsest nailing on their feet were situated nearest to the waterwheel and furthest from the vat. The finishing hammers with plain wooden heads were closest to the vat. This was done deliberately for the layouts of these mills are in reverse order to one another, and it must have been done to ensure a production flow of the pulp.

Papermaking

The techniques of making paper by hand that have survived in Ambert suggest a much earlier practice and origin than do similar mills in England. In both the Richard-de-Bas and Lagat mills, the original wooden vats have been replaced by circular ones made from copper. They have retained the 'pot hole' or 'pistolet' now fired by wood for heating the stuff. This gave so great a heat to the water that it was almost too hot to put in a hand on the coucher's side of the vat. A mechanical agitator is now fitted in the Richard-de-Bas mill but not at the Lagat so the following account concentrates on the latter mill.

The vat was prepared by the beaterman bringing a tub of pulp from the beater and ladling it in with a bucket. Size was poured in and, in recent times for making a particular type of paper, long white fibres were added to give a textured effect (a technique learnt from Japan). The stuff in the vat was then mixed with a paddle held in the hand and nothing else was added for the whole of the time while the next post was being made. This process was repeated for each post and the vat topped up with water if necessary.

When the vatman was ready to make a post, he gave the vat a further stir with the paddle. He then took a mould, put the deckle on top and scooped pulp upon it. Only about half the mould was dipped in to do this. He spread the pulp across, tipping some back into the vat over the far edge. At this point, the far edge of the mould might touch the stuff in the vat but the mould was never totally immersed. The shaking of the mould seemed to be minimal before it was placed, half over the edge of the vat, with the deckle still on top, resting on a stay of the bridge or draining-board. As the water drained out, some fell outside the vat into a drain on the floor.

The vatman paused, and then stirred the stuff in the vat with his right hand. At this point the coucher was couching the earlier sheet and when he had done this, he sent that mould to the far end of the bridge. Then the vatman passed his mould across to the coucher, lifting off the deckle with both hands at the edge nearest to him. The coucher took this mould and leant it against the asp (or stand) to drain. The vatman, meanwhile, was reaching for the empty mould to begin again. While the vatman was dipping, the coucher was laying the felt on top of the newly formed sheet of paper. While the vatman was stirring the stuff in the vat, the coucher couched the sheet. At the bottom of the post, the coucher rocked the mould from side to side a couple of times to couch the sheet, but as the post built up, he rocked it only once.

The press (a modern one built from steel girders with an electric pump operating an hydraulic ram above) was beside the vat. The post was pulled one yard (metre) across to it, triangular lengths of wood were placed on all four sides on top of the felts and then a series of heavy blocks of wood were placed on top, ending with a round one the same size as the head of the ram. Pressure was applied and then the 'weeping' edges of the felts were scraped with a piece of wood to remove the water. The pressure was increased and the edges scraped again to squeeze out the last bit of water. The ram was raised and the blocks of wood placed on the floor on the side of the press away from the vat. The post was lifted on top of them ready for laying off.

There was no layer, so the vatman and coucher placed a stool on either side of the post, one for the sheets of paper and the other for the felts. Two more stools were brought, on which the two men sat, facing each other. From his side, the vatman removed the sheets of paper, using both hands at the top corners, and laid them on a sloping stool, while the coucher lifted off the felts and turned them over on to the other stool. The pack or wad of wet sheets was carried by both men up to the drying loft in the top of the mill where the beaterman hung them up. The vat was refilled and work commenced again.

The lack of any storage chest for the stuff, the lack of any knotter and agitator, the continued use of the pot hole and stirring by hand, only two people at the vat and no layer, suggest that few innovations have been introduced to the handmade paper-mills in the Ambert region since 1700 and that here it is still possible to see, with the stampers and press at the Richard-de-Bas mill and the vat techniques at the Lagat mill, paper being made in the medieval tradition.

Notes

1. M.A. Peraudeau and E. Maget, *Le Moulin à Papier Richard-de-Bas* (Paris, 1973), p.10, but for another view see also Henri Gachet, 'Some Remarks and Reflexions Concerning the Introduction of Paper and Its Manufacture in the Mediterranean World' (Paper delivered to the International Paper History Congress, Fabriano, 1976).

2. M.A. Peraudeau and E. Maget, loc. cit., and E. Cottier, 'Le Papier D'Auvergne, l'histoire d'un vieux métier' (Volcans, 1974), but see also *Mémoires de l'Académie des Sciences, Belle-lettres et Arts de Clermont-Ferrand*, Tome XXXVII, 1937, Louis Apcher, 'Les Dupuy de la Grandrive; Leur Papeteries de la Grandrive et Barot; Leur Parent, l'intendant du Canada Claude Thomas Dupuy', p.124, 'Des cent trente-neuf moulins qui, en 1676, travaillaient dans le voisinage d'Ambert. . . .'

3. See illustration in the *Mémoires*, op. cit.

4. These observations are based on illustrations in the archive of E. Loeber who kindly allowed me to consult them.

5. *Mémoires*, XXXVII, pp.12–13.

6. See Livourne, *Encyclopaedia*, 1762, and J.J. Le F. de la Lande, *Description des Arts et Métiers, Art de Faire le Papier*, Vol. IV, 1761.

7. These observations are based on the recording work done by E. Loeber, now contained in his archives.

The Contributors

S.R. BROADBRIDGE (deceased) was formerly on the staff of the North Staffordshire Polytechnic.

STILLMAN DRAKE, who has devoted much of his life to Galileo, is Emeritus Professor of the History of Science, University of Toronto.

RICHARD L. HILLS is a Reader in the Department of History of Science and Technology at the University of Manchester Institute for Science and Industry and Director of the North-Western Museum of Science and Industry.

THOMAS P. HUGHES, biographer of Elmer Sperry, is Professor of the History of Technology at the University of Pennsylvania.

L.J. JONES is Senior Lecturer in mechanical engineering at the University of Melbourne, Victoria, Australia and has recently completed a study of the introduction of mechanical harvesting to Australia.

THORKILD SCHIØLER is interested in the mechanical and hydraulic technology of classical antiquity and is the author of *Roman and Islamic Water-Lifting Wheels*.

PROFESSOR D.G. TUCKER is Senior Fellow in the History of Technology at the University of Birmingham and a Member of Council of the Newcomen Society.

Contents of Former Volumes